THE HOUSING BOMB

THE HOUSING
BOMB

Why Our Addiction to Houses Is Destroying the Environment and Threatening Our Society

M. Nils Peterson, Tarla Rai Peterson,
and Jianguo Liu

The Johns Hopkins University Press
Baltimore

© 2013 The Johns Hopkins University Press
All rights reserved. Published 2013
Printed in the United States of America on acid-free paper
2   4   6   8   9   7   5   3   1

The Johns Hopkins University Press
2715 North Charles Street
Baltimore, Maryland 21218-4363
www.press.jhu.edu

Library of Congress Cataloging-in-Publication Data

Peterson, M. Nils, 1976–
The housing bomb : why our addiction to houses is destroying the environment and
threatening our society / M. Nils Peterson, Tarla Rai Peterson, Jianguo Liu.
pages cm
Includes bibliographical references and index.
ISBN-13: 978-1-4214-1065-4 (hardcover : alk. paper)
ISBN-13: 978-1-4214-1066-1 (electronic)
ISBN-10: 1-4214-1065-6 (hardcover : alk. paper)
ISBN-10: 1-4214-1066-4 (electronic)
1. Housing—Environmental aspects—United States.   2. Housing development—
Environmental aspects—United States.   3. Human ecology—United
States.   4. Urban ecology (Sociology)—United States.   5. Ecological houses—
United States.   6. Sustainable urban development—United States.   I. Peterson,
Tarla Rai.   II. Liu, Jianguo, 1963–   III. Title.
HD7293.P424 2013
363.5'40973—dc23   2012051093

A catalog record for this book is available from the British Library.

*Special discounts are available for bulk purchases of this book. For more information,
please contact Special Sales at 410-516-6936 or specialsales@press.jhu.edu.*

The Johns Hopkins University Press uses environmentally friendly book materials,
including recycled text paper that is composed of at least 30 percent post-consumer
waste, whenever possible.

CONTENTS

Many people contributed to this book, but we are most indebted to our family members, especially Shannon Peterson, who read endless drafts. Other family members who inspired us, gave us time to write, and provided valuable feedback include Gwen Peterson, Markus Peterson, Scott Peterson, Wayne Peterson, David Williams, Kenneth Powers, Qiuyun Wang, Julie Liu, and Jessica Liu.

Our colleagues in the Triangle Urban Long Term Research Area, the Center for Systems Integration and Sustainability, and the Human Dimensions of Fisheries, Wildlife, and Conservation Biology lab were instrumental in forming many ideas presented in this book. Students in the Communication Department at the University of Colorado, Boulder, asked hard questions about our suggestions for defusing the housing bomb. They pushed us to provide evidence that individuals can make a difference. Colleagues who offered ideas and constructive criticism include Cristián Alarcón, Danielle Endres, Jennie Stephens, and Elizabeth Wilson. Three anonymous reviewers provided crucial feedback and suggestions for improving the book. We received critical institutional support from Chris Moorman and Barry Goldfarb in the Department of Forestry & Environmental Resources at North Carolina State University, Michigan State University, the Environmental Communication Program at the Swedish University of Agricultural Sciences, and Texas A&M University. We thank all students and collaborators for many years of productive collaborations on research in the Wolong Nature Reserve, especially Zhiyun Ouyang of the Chinese Academy of Sciences and Hemin Zhang of the Wolong Nature Reserve, as well as Wolong residents for the opportunities to interview them.

Research presented in this book was funded by the National Science Foundation (Urban Long Term Research Area; Dynamics of Coupled Natural and

Human Systems; Partnership for International Research and Education; and Science, Technology, and Society Program), the National Aeronautics and Space Administration, the National Institutes of Health, the U.S. Forest Service (Urban Long Term Research Area), and the North Carolina Department of Environment and Natural Resources (Urban and Community Forestry Grants).

Finally, we must thank the Johns Hopkins University Press. Notably, Vincent Burke and copyeditor Kathleen Capels are as passionate about the subject as we are. They far exceeded our expectations in their enthusiasm for improving the manuscript as we revised it.

THE HOUSING BOMB

Introduction

The world is facing a housing bomb that will make the 2007 subprime mortgage crisis look trivial. Public attention focused on this housing bomb has been limited, with much greater emphasis having been placed on the environmental impacts of human population. Although the population problem has deep historical roots, Paul Ehrlich brought the idea to the masses in his ground-breaking book, *The Population Bomb* (1968), where he suggested that the human population was a bomb waiting to explode in an apocalyptic scenario of warfare, economic crashes, and mass starvation. He wrote, "In the 1970s, the world will undergo famines. Hundreds of millions of people are going to starve to death in spite of any crash programs embarked upon now." Fear generated by this prophecy gripped many among the emerging environmental movement. Fortunately, although political strife, poverty, and violence have combined with overpopulation to create regional famines, global collapse has not yet occurred.

This unexpected and fortuitous turn of events, however, can be explained by Ehrlich's own work. He argued that humans' impact on the environment was the product of human population, affluence (per capita consumption), and technology. When the human population exploded, technology and affluence prevented the predicted apocalyptic end of human society. Relatively affluent societies used their excess capital (natural resources, labor, knowledge, and money) to drive the green revolution and the ensuing biotech revolution, both of which dramatically increased crop yields and created the food surpluses needed to reduce hunger rates in the face of rapid population growth.[1]

In fact, environmentalists were faced with a paradox. While most indicators of ecological health have declined rapidly since 1960, most indicators of humans' quality of life have improved just as drastically.[2] Basic materials, health, and the freedom to choose all increased between 1960 and 2000. During this time period food supplies outpaced population growth, world hunger declined, and the average number of births per woman fell from 5 to 2.7. We do appear to be at a tipping point, however, because in 2008, both the percentage and the absolute number of hungry and malnourished individuals on the planet rose for the first time in recorded history.[3] The United Nations has revised their global population projections consistently downward over the last two decades. This decline in human population growth rates reflects a simple fact: people have curbed their addiction to having big families.[4]

Decisions to have a first child often revolve around psychological rationales (e.g., having a child to love), but decisions to have a second child, a step in avoiding rapid population declines, pertain to family building (e.g., having both a boy and girl).[5] Decisions about having three or more children typically involve considerations of the economic utility of children (e.g., helping parents with work or in their old age).[6] People might have seven children to provide a social-security net for themselves when none exists. They may have even more children when they know half will die from starvation before reaching adulthood. Brazil, a nation with nearly 200 million people, is a prime example of what happens when infant mortality and social security are addressed. The fertility rate in Brazil fell from 6.2 children in 1960 to 1.9 in 2008.[7] If immigrants are excluded from population calculations, every nation that has successfully provided food and social security for its citizens has a stable or declining human population. Some minority cultural and religious groups provide small-scale exceptions, but even Catholic Italy has had a fertility rate of less than 1.5 children per woman since the early 1980s, and that would lead to rapid population declines without immigration.

Unlike children, however, houses are an addiction.[8] While rising affluence and human well-being defused the population bomb, the same factors only whet our appetite for more houses, larger houses, houses on bigger pieces of land, and houses in beautiful natural landscapes. The factors that defused the population bomb built the housing bomb.

How Society Made the Housing Bomb

When people find social security and replace their investment in children with an investment in stuff, they invest in houses. Although population growth is stabilizing and even declining in many areas (such as regions of

New Zealand), the number of people per household is decreasing, and the number of houses is exploding.[9] The growth in housing numbers is faster than the increase in population in virtually every nation, irrespective of a country's development status. If the global number of households per capita eventually stabilizes at the current number in developed nations (0.4, or 2.5 people per household), 1 billion new houses would be needed, even with no additional population growth (see chapter 1).

The housing bomb is different from the population bomb, because human society has chosen to invest technology and excess capital in making the housing bomb bigger, rather than defusing it, as happened with the population bomb. People in wealthy countries don't invest their excess capital in making homes smaller, or in locating them in places with little environmental impact, and they far too rarely invest in making houses more efficient. Governments have also done their part to promote home ownership—and its accompanying sprawl—through the establishment of various fiscal and regulatory policies and institutions (see chapter 2). As the environmental and ecological disasters wrought by the housing bubble and the subsequent global economic collapse in 2007 and 2008 have demonstrated, affluence and technology will not rescue society from a housing bomb. Technology may help, but unless structural changes in society occur, affluence only makes the housing bomb bigger. Further, the social forces leading to declining rates of population growth (e.g., increasing wealth, and greater educational levels among women), reflect changes desired by humans on a global scale, but they currently contribute to an increasing proliferation of households.[10] Home proliferation poses a severe threat to the environment, and to the survival of a world that can continue to support human life.

Global trends toward larger houses, fewer people sharing housing units, and houses built helter-skelter across the landscape are driving a self-destructive binge of consumption that may bring global society to its knees. The housing bomb is a significant force behind wildlife extinction, dependence on fossil fuels (especially oil), unsustainable forest harvesting, abusive mining practices, climate change, water scarcity, and the loss of prime agricultural lands. Households serve as a reservoir into which resources are sunk, as a center where purchasing decisions are made, as a nexus for energy used in transportation and heating, and as the last tangible link most humans have with the land.[11] Current homes, and those being developed, are designed as resource vacuums. The typical infrastructure associated with houses sucks in resources through roads, water lines, gas lines, power lines, internet cables, and phone lines. Homes also come equipped with the means to excrete wastes, through sewer systems and trash-collection networks. This system makes households the nexus for resource consumption.

Houses start absorbing resources long before they come into existence. Home construction and its associated infrastructure play a primary role in the consumption of natural resources (e.g., oil, timber, copper, aluminum, iron for making steel, and the components in concrete). Once houses are established, they suck in even more resources. Homes also serve as point sources for pollution: waste spills into sewer or septic systems, lawn fertilizer washes into waterways, and motor vehicles generate emissions along veins of transportation (with their salted and oiled roadways) connecting households with employment centers. The bulk of global energy use is dictated by transportation to and from houses and by activities within homes (e.g., temperature control and lighting). The proliferation of households also limits the efficiency of energy use. Sprawling development requires more per capita energy for transportation and infrastructure. Declining numbers of people per household correlates with more overall energy usage per person, regardless of efficiency in other areas, including lighting, heating, cooling, and water consumption. With a fossil-fuel-based energy infrastructure, household dynamics drive climate change. The expansion of households into arable lands could cripple efforts to reduce the threat of climate change by using emergent agricultural products that provide sources of carbon-neutral energy.

Each home typically brings with it an entourage of unexpected effects. These include the three most important drivers of the current mass extinction of native plant and wildlife species: habitat loss, the introduction of exotic (non-native) plant and animal species, and landscape fragmentation.[12] The trend of households proliferating more rapidly in areas that receive more energy from the sun (e.g., migration to the Sun Belt, facilitated by air conditioning) encourages people to construct homes in areas that would otherwise have hosted a rich diversity of species. Of course, some people who either do not understand or don't care about the importance of biodiversity to human survival have no qualms about paving over wildlife habitat. But recent research on household proliferation also implicates people who prioritize protecting the environment. This group is more likely than others to situate their houses in places that endanger the very environments they value.[13] In the last five decades of the twentieth century, 28 million housing units were constructed adjacent to protected natural areas in the United States alone.[14]

House Addiction

The housing bomb's most pernicious attribute, however, isn't the environmental collapse it threatens, but the fact that it's driven by an addiction. Addictions reflect compulsive needs for an activity or a substance, despite knowing that it harms an individual's health, mental state, or social life. The obsession with huge houses located far from the crime and pollution of urban centers, and the use of them as a substitute for retirement savings, only seem reasonable because we are addicted to unsustainable housing.[15] Daniel McGinn referred to America's "houseaholism" as house lust.[16] This addiction creates the obvious problem of building the housing bomb, and a secondary problem of raising psychological barriers to defusing it.

As house addicts, we face a dilemma. The more we love nature, the more we want to live in it, but living in nature destroys it. All environmentalists want the clean air and water they fight for, but fewer want to live outside the natural environments they work hard to protect. Life in ultra-high-density housing far from the forests, wetlands, oceans, and deserts they love is simply not appealing. The dilemma becomes even more troubling if one assumes that experiencing nature in daily life is a prerequisite for being concerned about nature. In 1949 Aldo Leopold wrote: "A dead Chinaman is of little importance to us whose awareness of things Chinese is bounded by an occasional dish of chow mein. We grieve only for what we know. The erasure of Silphium [a wild sunflower] from western Dane County is no cause for grief if one knows it only as a name in a botany book."[17] The cultural isolation Leopold referred to has proven far easier to address than isolation from nature, because living with people from different cultures builds empathy and understanding. Living with nature builds empathy and a house. People living in natural habitats damage a piece of the land they are presumably learning to love. The common assumption that living in such areas builds concern for the environment may be overblown, however. Although research suggests that people building homes in or near spectacular natural areas are more concerned about nature, it also indicates that living in urban areas actually encourages more environmentally friendly behaviors.[18]

House addiction is a near-universal condition in America. We three authors join a long list of house addicts concerned about the environment. When we started this book we lived in quintessentially suburban towns (Cary, North Carolina; College Station, Texas; and Okemos, Michigan) within suburban-style neighborhoods; in houses near or above the national average in size

(2,000 square feet) having two, three, and four people living in each home; with average daily commutes of about 14 miles and heavily motor-vehicle-dependent lifestyles. We also clearly understood that the well-being of our families depended on fully functional natural ecosystems, ones that our addiction threatened.

House addicts are in great company. Some of the world's most beloved environmentalists had formative experiences, and even lived for most of their lives, in houses smack dab in the middle of nature. Throughout history, people who know how wonderful nature is have chosen houses in suburbia or exurbia. Gilbert White (1720–1793) is known as the founder of our modern respect for nature and, in many circles, as the seminal figure in the field of ecology.[19] One look at his countryside home should amply demonstrate the house addiction that has plagued nature lovers since at least the 1700s (fig. I.1). Charles Darwin couldn't handle the stress associated with living in London and sought refuge by retreating to the countryside. Henry David Thoreau's famous years on Walden Pond were spent in a suburban lakeside cabin that he built for himself.

Aldo Leopold—the father of wildlife conservation, forestry, and perhaps environmental ethics—bought a hobby farm (160 acres) near Baraboo, Wisconsin, in 1935, the same year he founded the Wilderness Society. Leopold and his family renovated an abandoned chicken coop on the farm to use as a weekend home. The now-famous Shack was a far cry from the homes used as weekend getaways by many today, but Leopold's writings imply that he loved living in his second home / shack in the countryside, and that doing so played a formative role in his environmental views.[20] James Lovelock developed the Gaia hypothesis, which suggests viewing Earth as a single organism.[21] He became somewhat of a celebrity among environmentalists who had a penchant to anthropomorphize (attribute human attributes to) Earth. To explain why he moved to a rural home in Wiltshire, England, Lovelock asked, "How can we revere the living world if we can no longer hear the bird song through the noise of traffic, or smell the sweetness of fresh air?"[22] When agribusiness moved into Wiltshire, Lovelock relocated even farther away from the city, to Bowerchalke, where he ironically hoped to escape the destruction of the English countryside.

Renowned nature writer and environmental activist Terry Tempest Williams split her time between living in the shadow of Arches National Park and Grand Teton National Park (www.coyoteclan.com/bio.html) during the period when our book was being written. Al Gore, a former vice president of

Figure I.1 Gilbert White's countryside home. Courtesy Ludi Ling.

the United States, was taken to task for living in a 10,000-square-foot house in the suburbs of Nashville that consumed 12 times more electricity than a typical Nashville home that was four times smaller in size. That glaring problem was all the more conspicuous when set alongside Gore's recent Academy Award for *An Inconvenient Truth*, a documentary film about the dangers of climate change. Gore tried to pay for his house addiction with carbon offsets and solar panels, but before he finished reconciling this first mansion with his concerns about energy use, he bought another with 6,500 square feet of living space and got divorced, halving the number of people actually living in his houses.

These famous environmentalists are not unique in their apparent desire to live in environmentally damaging houses. Our recent research with a community near Yellowstone National Park suggests that environmentally oriented people may actually seek out rural, natural, and ecologically sensitive areas in which to build or buy their homes.[23] Environmentalists may be a minority in most societies, but reverse migration (from urban to rural areas) and rapid suburbanization have characterized nations throughout the developed world during the last two decades. Many non-environmentalists decided that they didn't like raising children in dense urban centers decades ago. U.S. census data started showing rings of population growth, made up of young families, surrounding metropolitan areas like Baltimore, Atlanta, and Chicago before 1990.

These families had good reasons to abandon the city. Several studies have found that connection to outdoor spaces can help children cope with attention deficit disorder, combat obesity, decrease instances of asthma and myopia, reduce stress, and improve mental health.[24] These physical and mental health issues and their side effects typically persist far past childhood. Diverse and poorly understood causes of obesity range from ethnicity to the price of vegetables, but where we put our houses has also proven to be important.[25] Research based on the National Longitudinal Study of Adolescent Health's 1994–1995 baseline data examined the differences in U.S. adolescents' risk of obesity and in their physical activity patterns according to neighborhood characteristics.[26] Study participants were grouped into six categories: (1) rural working-class; (2) exurban (rural suburbs); (3) newer suburban; (4) upper-middle-class, older suburban; (5) mixed-ethnicity urban; and (6) low socioeconomic status, inner-city areas. Children living in rural working-class, exurban, and mixed-ethnicity urban areas were approximately 30% more likely to be overweight than children living in newer suburbs, independent of each individual's socioeconomic status, age, and ethnicity. The paradox is that these health benefits come from increasing sprawl. The housing bomb presents a unique challenge, because the people most committed to defusing this bomb are often implicated in building it. The desire to have homes constructed near natural areas and far from the problems associated with urban decay only exacerbates this challenge.

Isolation from Nature

Ironically, the wish to reconnect with nature is contributing to the housing bomb. Mounting public concern about children's isolation from nature has created a powerful force driving the expanding suburban development that gobbles up natural areas. Human isolation from nature started when our earliest ancestors climbed out of the trees, but accelerating social change in recent decades has highlighted the phenomenon and supported a backlash against it. Richard Louv's book, *Last Child in the Woods*, popularized the idea that modernization and suburbanization isolated people from nature in the past generation,[27] but trading backyard tree houses for video games during the last half of the twentieth century probably had far less impact on humans than trading primary dwellings in trees for caves had in human evolutionary history. Humans have been working to isolate themselves from nature since our species evolved. Tools, shelter, and clothing all provide ways to insulate ourselves from nature in the raw. Nature unmediated by technology, tools, or medicine is painful and unpleasant for most people. Even the most extreme back-to-nature survivalists utilize a suite of devices to distance themselves from nature, including clothing, a knife, and equipment to start a fire. Simply put, without some degree of separation from nature, most humans could not survive at all.

The idea that humanity has lost something valuable through its progressive isolation from nature is an ancient one. Plato's (400 BC) allegory of the cave begins by using the example of a group of people who, having spent their lives in a cave, would mistake shadows cast on the wall for reality. He suggested that the human community exists inside a cave formed by its isolation from nature.[28] This symbolic cave represents the biases of subjectivity, values, and politics. A philosopher could leave the cave and apprehend reality in nature. The philosopher then had an obligation to the society remaining within the cave. Philosopher kings, who freed themselves from the cave (from politics, values, and subjectivity in society) to find truth in nature, would rule over a good society. They would bring their newly discovered truth, unaltered by human perceptions, back into the cave, in order to silence the "endless chatter of the ignorant mob."[29] Modern society's faith in objective science suggests that the division between reality in nature and subjectivity in society has defined Western society for centuries.[30]

The limited level of isolation from nature experienced by the Greeks persisted in multiple agrarian societies until the industrial revolution allowed most people to safely ignore a suite of natural forces crucial to farming. Farmers must understand hydrology, nutrient cycles, soil productivity, erosion, and (in some cases) animal nutrition and population dynamics. Agricultural work does not guarantee stewardship, but it does promote some physical interaction with and awareness of nature. The industrial revolution started moving the world's population off farms and into urban areas. Urbanization is rapid in developing nations, such as China, where small-scale farmers are being lured into urban areas to take manufacturing jobs, and it is almost complete in developed countries, where more than 80% of their residents now live in urban areas. We do not want to romanticize agricultural lifestyles, but the transition to an urban society has undeniably severed human connections to nature. Farmers must know where milk, eggs, chickens, rice, and sweet potatoes come from, and their lives revolve around rainfall and the seasons. Residents of inner-city neighborhoods can safely ignore these issues, at least in the short term.

The recent concern about isolation from nature may, more than anything else, reflect a nostalgia-fueled backlash against accelerating social change. Globalization's many forms have rapidly pushed forward changes in how humans live their daily lives. The threats posed by terrorist attacks in the United States in 2001, and the ensuing war on terror, have further supported nostalgia for how the world was. These contexts make appeals to the way things used to be particularly effective, whether they relate to children building tree houses or people buying food at local produce stands. They also made news about children having disorders from playing video games, and people growing sick from eating fast food instead of cooking their own meals, particularly alarming.[31] Every naturalist since Gilbert White, in the eighteenth century, has extolled the importance of exposure to nature, and the Tbilisi Declaration of 1977 galvanized environmental educators, who worked to create environmentally responsible behaviors in society by helping students make connections between themselves and nature.

How to Defuse the Housing Bomb

To defuse the housing bomb, we will need ways to connect with nature other than building houses in it. Two strategies have emerged. The first is to use both social and material engineering to satiate house addiction in urban cores. The second is to address the problem with creative combinations of clustered housing and shared open space in suburbia, thus reducing environmental impacts.

We learned about the first alternative while working to save endangered species from losing vital habitat to housing developments in the Florida Keys. The miniature Key deer and a host of plant and animal species on the islands and in the oceans around the islands were and are threatened by housing developments and their associated landscape changes (e.g., people suppressing natural wildfires, squashing animals with their motor vehicles, and killing coral reefs because of leaky septic systems). The conservation of Key deer sparked violent conflicts between supporters of housing development and advocates of land preservation.[32] During our research we met Joel Rosenblatt, an engineer who gave us a signed copy of his book, *Space on Earth: The Story of the Urban Mountain* (1996). Rosenblatt assured us that the book contained the secret to convincing people to live in dense urban centers. Rosenblatt himself, however, lived on Summerland Key. The island was basically a nature resort, equipped with its own airstrip and a canal system for the residents' boats. The natural habitat on the island had long since been replaced by houses, but residents could step out of their homes onto boats sitting in their personal canals—which were dredged through sensitive wetlands—and zip over to a beautiful coral reef or small mangrove island. Reading *Space on Earth*, one cannot help but think that Rosenblatt's urban mountain would strike terror into the hearts of most Summerland Key residents. The urban mountain was a giant box that housed 50,000 people on 30 acres of land. Inside of the box were intricate interchangeable parts capable of simulating all the suburban amenities people wanted. Synthetic homes, condominiums, streets, and parks could be moved and transformed by planners as if they were playing with LEGOs. The city could operate with a fraction of the energy demands, land requirements, and waste outputs of traditional human settlements. With the exception of farmers, no one needed to leave the box. Blissful residents could cavort among the painted concrete trees, just like visitors at Disneyland. Unlike Disneyland, however, Rosenblatt's urban mountain sparked little interest among consumers.

David Owen's book, *Green Metropolis* (2009), highlights the fact that high-density urban centers with people living in small high-rise apartment buildings can achieve the same effect as an urban-mountain-type development. Owen, however, turns out to be a house addict, too. In the first pages of the book he bemoans the ecological catastrophe wrought by his family moving from a 700-square-foot Manhattan apartment to a three-story rural house next to a nature reserve where bears and turkeys roam in his yard. He implies that he would move to a small Manhattan apartment from his rural, three-story, energy-gobbling Connecticut home if he didn't believe that someone else would simply take his place. This justification for high-impact homes arises from what is known as the tragedy of the commons. Garret Hardin popularized the tragedy of the commons idea with his account of herders sharing a common meadow.[33] In this parable, each of the herders will choose to overgraze the meadow and destroy their own livelihood, because they all quite rationally assume that their compatriots will do so if they don't. Owen and other environmentally conscious people with dream homes in the woods can rationalize their choices by assuming that someone else would take over their lifestyle if they settle for a small steel-and-glass box in Manhattan or a simulacrum of nature in an urban mountain. Invoking the tragedy of the commons, however, is tantamount to admitting that far too many people will never accept high-rise living. So what are the alternatives to forcing people into urban mountains?

Various forms of conservation developments provide the leading alternatives to living in densely populated urban centers. (These alternatives are discussed in chapter 6.) Unfortunately, the fact that they are merely alternatives to living in dense population centers belies their flaws. Conservation development, popularized by Randall Arendt, advocates various permutations of clustered housing and shared open space. Carefully implemented conservation development offers an improvement over traditional suburban sprawl, but the conservation benefits associated with moving people from traditional suburbs to conservation suburbs are dwarfed by the benefits of moving people from suburbs to urban mountains like New York, London, or Tokyo.[34]

Conclusion

There is no magic pill to cure house addiction, so society has tried to have its cake and eat it, too. Society has adopted two mutually incompatible strategies—building houses within nature, and separating nature from houses with protected areas—in a race to the bottom (when individuals seeking optimal results create suboptimal outcomes for society) for land use. In 2010, there were over 155,000 terrestrial protected areas (www.protectedplanet.net), covering over 13% of Earth's land surface.[35] This race to the bottom has proven to be horribly unjust, with many of the world's poorest people being expelled from their homes in newly established protected areas at the same time as the world's wealthiest people build homes inside and beside these areas.[36]

The housing bomb is a complex problem that defies simple descriptions and solutions, but the recent global recession, which is tied to housing, has created both unprecedented need and an opportunity to address our addiction. The housing crash that started in 2005 ended the largest real estate boom in U.S. history. The housing boom behind the bubble was global, and it was environmentally detrimental, both because houses were multiplying far faster than people in almost every country, and because homes were multiplying fastest in environmentally sensitive areas.[37] Nations with biodiversity hotspots (areas with substantial numbers of native plant and animal species) were experiencing the fastest growth in the number of new houses. In the United States, nearly 1 million homes were built inside national forests, and 26 million houses were constructed within 50 kilometers (31 miles) of protected areas between 1940 and 2000.[38] Had the housing bubble failed to collapse, by 2030 previous trends would have placed an additional 1 million homes within 1 kilometer (2.2 miles) of protected areas and 17 million homes within 50 kilometers of protected areas. In the United States, house addiction reached its zenith in 2005, when 40% of all homes that were sold were second homes.[39] Almost every other home that was sold was to a buyer who didn't even want to live in it full time. Not only did the houses have a large geographic footprint cut from prime agricultural land and natural areas, but travel to the homes created a huge carbon footprint. Nearly a third of all the vacation homes sold were more than 1,000 miles from the buyer's primary residence.[40]

Although the catastrophic environmental effects of housing have proven to be a weak tool for motivating social change, the subprime mortgage crisis, the ensuing credit crisis, and the eventual meltdown of the global economy in 2008 were harder to ignore. In the United States, the Federal Reserve reported that households lost almost 20% of their net worth, totaling $11 trillion in 2008. This meltdown actually brought some attention to problems associated with a loosely regulated system designed to churn out homes as fast as humanly possible. So, how can we defuse the housing bomb? In the short term, we can remodel houses and the way we distribute them on the landscape, so as to reduce environmental impacts. Over the long term, we need to find some way to interact with nature besides building a house in it, and to find some means to accumulate wealth beside investing in houses.

Solutions to the housing bomb include feel-good suggestions, such as renovating urban cores, making suburbia more environmentally friendly, and redirecting house addiction into a green consumption (choices based on environmental criteria) of household-related amenities. Unfortunately, any real solution will include some less popular changes, such as forcing people to pay the full social and environmental costs for choosing to live in suburbia or in natural areas. The Apollo 8 astronauts took an iconic photograph in 1968: an image of Earth rising above the moon's horizon. This image helped catalyze the global environmental movement because, for many people, it encapsulated the fragile and finite nature of humanity's home in the universe. This book offers a renewed focus and new perspectives on the fundamental roles of *home*.

The volume is broken into three sections: documenting how the housing bomb has emerged, describing how it has evolved (using two case studies), and suggesting how it can be defused. Chapter 1 discusses the social, economic, and historical factors that have influenced changes in the number of households per capita over the last several centuries. Chapter 2 explains how home ownership was envisioned as a way to create human freedom, but instead is evolving into one of the primary mechanisms for taking that freedom away. Chapter 3 uses a case study to describe how the housing bomb threatens the Greater Yellowstone Ecosystem, and how one community has fought to defuse it. Chapter 4 chooses another case study to document how the housing bomb threatens the heart of giant panda habitat, and describes both successful and unsuccessful attempts to address the crisis. Chapter 5 illustrates how readers can defuse the housing bomb by making easy and economical changes to their own homes. Chapter 6 enumerates key policies that can be used to address the housing bomb at household, neighborhood, and city scales; chapter 7 tackles the same issue at national and international scales.

CHAPTER ONE

Household Dynamics and Their Contribution to the Housing Bomb

Unsustainable patterns in human relationships with the Earth fill the literature propagated by virtually every environmental science discipline and every environmental organization. Changing these patterns requires altering household dynamics (making temporal changes in household numbers, types, and locations). Household dynamics may have already supplanted (or soon will supplant) population growth as the primary environmental threat posed by humanity.

The relationship between human population size and natural resource consumption has concerned scientists since 1798, when Thomas Malthus published his seminal work, *An Essay on the Principle of Population.*[1] In the 1970s Paul Ehrlich helped frame perceptions of the influence of overpopulation on the environment with his model I = PAT, hypothesizing that impact (I) = population (P) × affluence (A) × technology (T).[2] Since the early 1970s, research has consistently supported the idea that population size determines resource use and environmental impacts in multiple areas, ranging from water use, to pollution, to plant and wildlife extinction.[3] Population growth rates, however, are slowing, without concomitant declines in negative environmental impacts.

Throughout most of the world, population growth has decreased as the human condition improves. The opposite relationship holds true for households: as the human condition improves, households proliferate. During the second half of the twentieth century, fundamental changes in our culture, economics, education, and technology combined to drastically alter household dynamics. In an ironic twist on the attempt to moderate the rate of population increases, aging has led to drastic increases in the number of households per capita.[4] Other cultural phenomena, including increasing divorce rates and decreasing incidences of multigenerational households, also contribute to household proliferation.[5] Even though population growth has been flattening out, human relationships with nature are becoming progressively more destructive.[6]

A growing body of research suggests that "households" can supplant "population" in the I = PAT model, and average household size has an even larger effect on resource consumption and biodiversity than does population.[7] Similarly, households appear to better predict amounts of carbon dioxide (CO_2) emissions, fuelwood consumption, per capita use of automobiles, and species endangerment.[8] Energy consumption increased by 2.1% annually in the 1970s and 1980s, and population growth only accounted for 0.6% of that amount; the remainder (three times more) was related to per capita increases in energy use propelled largely by households.[9] These findings are alarming, because while population growth has been a problem that is coming into check, the growth in household numbers is not. The number of people living in each house (household size) is declining around the world, and it appears to be declining fastest in biodiversity hotspots (places with high numbers of rare and unique animal and plant species).[10]

Households are the primary CO_2 emitters, and they are the basic socioeconomic units of resource consumption.[11] In the United States, direct and indirect energy consumption by households makes up 85% of total energy use.[12] U.S. households directly contribute almost 40% of the nation's carbon emissions, which is higher than the percentage of emissions from the entire industrial sector in the United States, as well as the amount of emissions from any other country except China.[13] In India, households use over 70% of that nation's total primary energy.[14] Thus households will significantly shape global CO_2 emissions in the future.[15]

Despite the implications of global declines in household size, far more research focuses on population dynamics. Although most high school social studies courses teach their students about demographic transition theory as it applies to population, few address whether there is a demographic transition theory for household dynamics. In the case of population, extensive research has documented how factors such as the education of women, decreasing infant mortality, and urbanization are able to drive down the rate of population growth to replacement levels (2.1 children per family) or even lower across diverse cultures and nations. The possibility that similar factors could drive down household sizes to some as-yet-unknown size is both intriguing and important to explore.

Few regional and national studies have focused on household dynamics, and virtually none have examined historical patterns in household dynamics and their relationship with the environment.[16] The little information that does exist is limited to particular cases, villages, or individual nations.[17] Although household dynamics offer great potential for developing more sustainable human societies, they represent a double-edged sword. Changes in how many people share a home, where houses are built in relation to each other and to key features of ecosystems, the size of homes, what materials are used to build homes, and what purposes homes are built for can both solve and exacerbate today's environmental challenges.

Housing has played a central role in the sustainability of human societies since their beginning. Unfortunately, very little attention was paid to global-scale household dynamics prior to 1900. Given that household size has been found to directly affect natural resource consumption, and that this consumption is threatening sustainability in many parts of the world, household dynamics is a topic that deserves more attention. In this chapter we describe the history of household dynamics around the world, document how these dynamics have impacted the environment, and highlight several important trends that will shape efforts to address the housing bomb.

Historical Household Dynamics

We collected data for household dynamics since 1990 from previous research.[18] Historical data were collected from government documents (censuses, statistical abstracts, yearbooks, and books authored by demographers) and the United Nations' online listing of Population and Housing Censuses (http://unstats.un.org/unsd/demographic/sources/census/censusdates.htm). The distinction between developed and developing countries was based on an informal classification used by the United Nations' Statistics Division, in which Japan, Australia, New Zealand, Canada, the United States, and Europe are termed "developed," and the rest of the world is called "developing."[19]

A demographic transition in household dynamics occurs where the average household size tends to start at around 5 individuals per household and to decrease to just above 2.5 individuals after nations urbanize and industrialize (fig. 1.1). During periods for which records are available, household size decreased in the following 19 countries and territories: the United States, England and Wales, Ireland, Canada, Luxembourg, New Zealand, Brazil, France, Japan, Australia, Hungary, Mexico, Trinidad and Tobago, Greece, Puerto Rico, Singapore, China, Egypt, and the Netherlands.

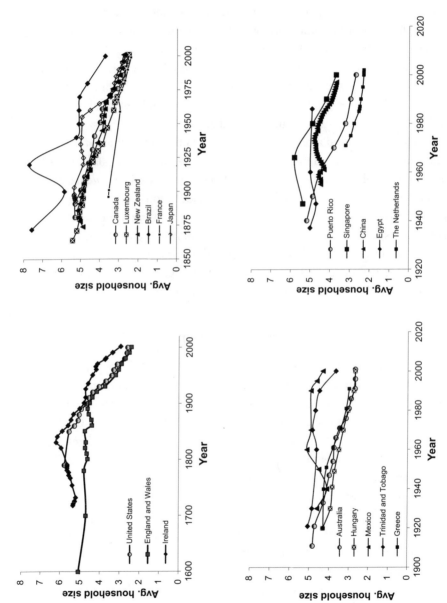

Figure 1.1 People living in a housing unit over time, for 19 nations. Bradbury, Peterson, and Liu (in review).

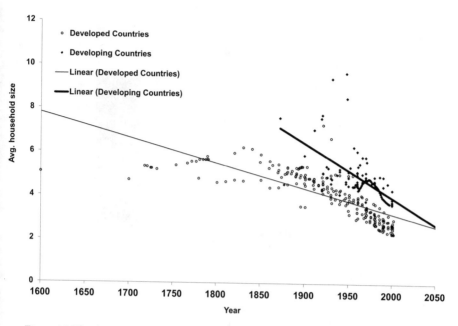

Figure 1.2 The decreasing number of people living in a housing unit in developed and developing countries.

The graphs do not include countries with fewer than four data points regarding household size or those without data for years prior to 1975. In 1947, Singapore registered the largest average household size in our dataset, at 9.66 individuals. The lowest was 2.29 for the Netherlands in 2002. A comparison of developed (*n* =12) and developing (*n* = 25) nations suggests that household size is falling more than twice as fast in developing nations. If we base our assumptions on previous and current trends, these data suggest that household sizes in developing and developed nations will converge at around 2.5 people per household by 2050. Although declines in household size seem to be linked over time to urbanization and industrialization in developed nations, declines occurred without major economic development in developing nations. All nations experienced a net decrease in average household size over time, but some (such as Brazil, Mexico, Trinidad and Tobago, China, and Egypt) fluctuated or did not decrease steadily. The demographic transition in household size appears to be more erratic and abrupt in developing nations (fig. 1.2).[20]

Household numbers increased faster than population growth in every na-

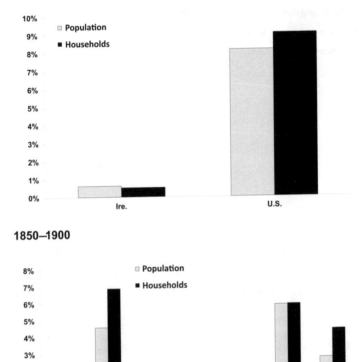

Figure 1.3 The percentage increase in population and households in the 1800s. Bradbury, Peterson, and Liu (in review).

tion and in every time period since 1850, with the exception of Ireland (figs. 1.3 and 1.4). During the 1700s and through the great famine in the mid-1800s, Ireland's people were exploited and poverty engulfed the nation.[21] This may explain the population in that country growing faster than its household numbers in the early 1800s (fig. 1.3, 1800–1850). Similarly, when Ireland experienced a population decline after the famine of the mid-1800s (fig. 1.3, 1850–1900), its population size decreased faster than the number of households. Thus household numbers appear to be more resistant than population to some negative demographic drivers, such as famine and emigration.

1900—1950

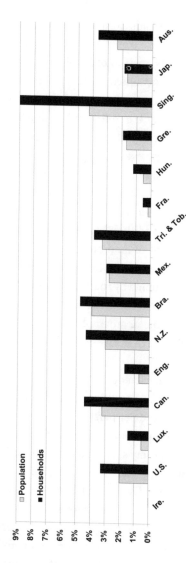

1950—2000

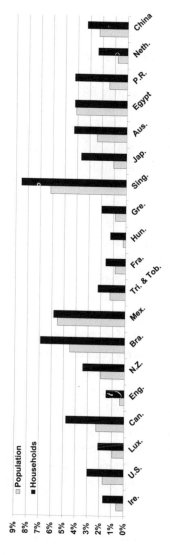

Figure 1.4 The percentage increase in population and households in the 1900s. Bradbury, Peterson, and Liu (in review).

Historical trends in household size suggest that major adaptations are needed to cope with increasing household numbers, even if human populations begin to decline. If the average household size had been 2.5 people globally in 2010, then the number of households would have been 41% higher. That equates to 800 million additional households in the 172 countries with available data (2.7 billion households, instead of 1.9 billion households).[22] If each of the additional households occupied a 232-square-meter (2,500-square-foot) home (the average U.S. house size in 2012),[23] then an additional 185,806 square kilometers (71,740 square miles) of housing area would be needed. If each house halved its footprint by being two stories high, then 92,903 square kilometers (35,870 square miles) of land area would still be needed for housing. That equates to an area twice the size of California. Even if the size of an average house globally is 50% or 25% that of an average U.S. house, 46,451 and 23,226 square kilometers (17,935 and 8,968 square miles), respectively, would be needed for the additional households. These estimates are conservative, because the amount of land area used for other purposes (e.g., roads, yards, and gardens) is not included. Some research suggests that two to four times as much land is occupied by roads and infrastructure designed to service each home as is tied up in the actual amount of land used for the home.[24] If land area is included, the demographic transition to 2.5 people per house globally could require an area nearly half the size of the continental United States.

If these statistics are not staggering enough, each of those houses will demand more household products and furniture, in addition to land. Households would be less efficient in their resource use per person, because fewer people share goods and services in smaller households.[25] If the global trend toward household sizes of 2.5 individuals continues, then at least 800 million additional televisions, refrigerators, washing machines, radios, air conditioners, and other types of durable goods would be needed, assuming each household has one of each product, even if no additional people are added to population levels. The amount of energy required to produce and operate the appliances is enormous. The implications for the amount of energy that would be used to transport people between these houses, their work, and recreational areas adds another staggering amount to the energy demand associated with the ongoing demographic transition in household size. The repercussions will be most acute in places where transportation relies on private motor vehicles.

Drivers of Decreasing Household Size
Declining Fertility

The global trend of declining household sizes, from over 5 individuals per household to approximately 2.5, has many social and cultural drivers, including industrialization, urbanization, declining fertility, affluence, divorce, and aging. Household dynamics, however, defy simple explanations. For instance, the industrial revolution, urbanization, and changes in fertility rates occurred about 100 years earlier in the United Kingdom than they did in the United States, yet rapid declines in household size started in the early 1900s in both nations (fig. 1.1). Despite the complexity and unique nature of household dynamics across nations, some generalizations can still be made. First, the number of children in a family is a factor, and declining fertility rates contribute to declining household sizes. Since 1950, fertility rates have fallen from 4.9 children per woman to 2.6 globally.[26] They fell by 30%–50% in developed countries: from 3.7 children per woman to 2 in North America, and from 3.1 to 1.8 in Europe and Central Asia. Declines were even greater for East Asia (from 5.6 to 1.8), Latin America (from 6 to 2.3), the Middle East and North Africa (from 6.9 to 2.7), and South Asia (from 6 to 2.8). Sub-Saharan Africa was notable for being the only region in the world where fertility rates did not drop drastically (merely from 6.6 to 5) during the last half of the twentieth century. As fertility rates go down, household sizes drop. Declining fertility can be linked to modern contraception, the education and emancipation of women, lower infant mortality, and the advent of social security as standards of living increase globally, but whatever the cause, household size shrinks.[27] Declining fertility rates can help decrease the number of households by reducing the population over long periods of time, but in the short term, the number of houses goes up. This is not true, however, for other factors, such as those discussed below. These are major contributors to household proliferation, and they do not cause population declines.

Aging

Household sizes have continued to decline rapidly, even in developed countries, where fertility rates have been stable for decades. Aging provides one explanation for this phenomenon. Based on data from the United Nations, we found that in the year 2000, households with an elderly member had between 1.3 and 3.9 fewer people than houses without an elderly resident.[28] This pattern persisted across national, continental, regional, and global levels. Globally, households without elderly residents had 2.6 more people than those with elderly members. The age-related difference in household size was smaller in more-developed regions (1.4 individuals) than in less-developed regions (3.1 individuals). At the continental level, the difference was smallest in North America and Europe, and largest in Africa. Major factors behind the decreases in household size that are associated with aging include husbands dying sooner than wives, and fewer multigenerational households.[29]

Devising sustainable housing solutions for the exploding population of people facing the prospect of aging while living alone can provide a win-win scenario by reducing the ecological footprint of housing while improving the quality of life for the elderly. Aggressive efforts to provide housing with separate living quarters but shared areas for eating and socializing can improve the quality of life for older singletons, as well as reduce the threat of millions aging alone in detached, single-unit suburban housing.[30]

Independent Life Stage

While older persons contribute to the growth in housing numbers by living by themselves for longer periods of time, the younger generation is adding its share by leaving home sooner. This assertion may seem counterintuitive, given the deluge of media commentaries about the boomerang generation created by the economic woes of recent recessions. The recent spike in adult children returning home (3%–4%), however, represents a blip in the overall trend of children leaving home as young adults in ever-increasing numbers (fig. 1.5). In the 1940s, over 70% of the unmarried young adults in the United States lived with their parents; in 2000 that figure had dropped by 35%. The independent life stage (where young unmarried adults live independently from their parents) represents a novel demographic development that started in the late 1940s, and it contributes to millions of additional households in the United States.

The percentage of young adults living with their parents increased slightly in the early 1900s, primarily because parents were living longer, so their adult children had the option to live with them.[31] After the 1940s the number of independent unmarried householders doubled, even though the lifespan of parents increased further. This shift can be linked to rising affluence and a nearly 500% increase in the number of young adults attending college. The

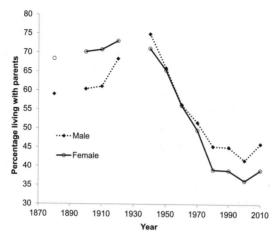

Figure 1.5 The percentage (since 1880) of unmarried young adults (age 20–29) in the United States living with their parents.

12% of young adults (those in their 20s) who had attended college in 1940 increased linearly to 54% of the men and 63% of the women by 2000. Since more than 50% of the 48 million people in their 20s were not married in 2010, and almost 70% lived independently from their parents, the independent life stage has created new households for approximately 17 million young unmarried adults. At an average household size of 2.5 people, that equates to at least 6.8 million households. If a similar independent life stage emerges in developing countries, that would create 160 million households composed of unmarried people in their 20s.

Divorce

Divorce is increasingly common globally and has huge impacts on the formation of households and average household size. In the United States, there were 15.6 million divorced households (households headed by a divorced person) in 2000, which equated to about 15% of all U.S. households.[32] This relatively high percentage of divorced households persists in the population, despite a constant turnover between new divorces and an almost 75% remarriage rate.[33] Between 1998 and 2002, household size in divorced households was 27%–41% smaller than in married households (households headed by a married person) in 12 different nations: Belarus, Brazil, Cambodia, Costa Rica, Ecuador, Greece, Kenya, Mexico, Romania, South Africa, Spain, and the United States.[34] Since a divorce splits a single household into two or more households, combining divorced households to have the same average household size as married households would lead to 7.4 million fewer households in these countries alone.[35] Many factors affect divorce rates,[36] but unemployment may be one of the most important. Many anecdotes have suggested that divorce rates decrease during recessions. Recent research provided empirical evidence for these observations by finding a significant and robust negative relationship between unemployment and divorce rates.[37] A 1% increase in the unemployment rate was associated with a 1% decline in the divorce rate in the United States between 1976 and 2009. This relationship thus highlights another way something society values—prosperity—contributes to the housing bomb.

Drivers of Increasing Environmental Impacts from Housing
Growing Physical Size of Homes

Household dynamics shape sustainability through the physical size and location of houses, as well as their overall numbers. The physical size of homes is increasing around the world. In the United States, the average house size grew from slightly over 90 square meters (1,000 square feet) in 1950 to almost 210 square meters (2,200 square feet) in 2002.[38] During 1978–2002, per capita floor space increased from 8.1 to 26.5 square meters (87–285 square feet) in rural China, and from 6.7 to 22.8 square meters (72–245 square feet) in China's urban areas.[39] The size of U.S. homes doubled in 50 years, but the size of Chinese houses more than tripled in half that time. One can hardly begrudge the rapid increase in home size in China, given that the per capita square footage of Chinese houses in 1978 was a scant 20 square feet more than that allocated to residents in the Texas prison system the same year.[40] This finding, however, suggests that rapid increases in the physical size of homes will increase the potency of the housing bomb, since the number of households has already exploded from the demographic transition toward 2.5 people per household.

Sprawl

Sprawl makes the trends of fewer people sharing a home and houses growing physically larger even worse for both the environment and humans' well-being. Urban sprawl can be difficult to define, but it generally refers to patterns of low-density development adjacent to metropolitan areas.[41] The label often implies that the overall area of land being developed is growing at a faster rate than the population.[42] Sprawl was pioneered as a part of urbanization in the United States, and this trend has now spread to other nations.[43] By its very definition, sprawl has wide-ranging impacts on the environment and on humans' well-being, and the statistics are mind boggling. During the 1990s, over 15% of the private land in the United States was developed.[44] Farmland was being converted to housing sites at a rate of more than 1,000,000 acres per year in the 1990s, rates that would completely eliminate agriculture in five decades, even in agricultural states like Florida.[45] Over half of the agricultural acres used for dairy, fruit, and vegetable farming face development pressure. The amount of occupied land in many major metropolitan areas grew between 4 and 10 times faster than the population from 1960 to 1990.[46] By 2005 the turfgrass lawns in sprawling suburbia constituted more than 40,475,861 acres in the United States, an area 3 times larger than that dedicated to corn.[47] Lawn acreage is expanding annually, with 23% of new urban lands (1,667,961 acres per year) devoted to turfgrass.[48]

The massive changes in the landscape wrought by sprawl have created equally massive impacts on the environment. The production of turfgrass alone has shifted urban biogeochemical cycling and the global carbon cycle.[49] Maintenance of sprawling suburban landscapes contributes to environmental degradation through the use of chemicals, including fertilizers, pesticides, and herbicides, which negatively impact water and soil quality; the expanding use of automobiles and lawn mowers, which contribute to the increased carbon dioxide emissions that are linked to global climate change; and irrigation, which threatens limited water supplies.[50] Runoff from suburban homes is the leading cause of water pollution, and suburban homes use more energy than urban homes of equal size and age, because of increased automobile use and driving times in suburban areas.[51] Suburbanites use 12 times as much gasoline as residents of urban cores, explaining why the United States contributes more per capita to climate change than any other nation.[52] Motor vehicle emissions account for about half of all toxic air-pollutant emissions (carbon monoxide, nitrogen oxides, volatile organic compounds, and benzene).[53] Further, typical landscaping in suburbia tends to be relatively sterile in terms of wildlife habitat, as it lacks the vertical and horizontal structure and the native plant species wildlife require for food, cover (e.g., places to hide), and reproduction.[54] The impervious surfaces (e.g., pavement) associated with sprawl in the region around the city of Atlanta prevent between 60 and 130 billion gallons of rainwater from replenishing the aquifers.[55]

The environmental effects of sprawl typically reduce humans' quality of life. The open space that is replaced by sprawl is more than an amenity; it provides key ecosystem services needed for humans' health and well-being. Forests and wetlands purify air; filter water; and produce food, fiber, and fuel used by humans. For example, the loss of mangrove forests and wetlands around New Orleans contributed to billions of dollars in property damage and the tragic loss of human life after Hurricane Katrina. Sprawl also threatens food security by supplanting productive farmland.

The nitrogen oxides produced by cars (32%) interact with sunlight to create ground-level ozone, which burns the inside of human lungs when inhaled. Nearly every region with plenty of sunlight (lower latitudes during the summer) and lots of cars has ground-level ozone problems that cause asthma, chronic bronchitis, reduced lung function, and premature death. Research suggests that motor vehicle emissions in Europe account for twice as many premature deaths as car accidents.[56] Children are the most sensitive to lung damage from these pollutants. Ground-level ozone also hurts plants and reduces crop productivity.[57]

In addition, sprawl contributes to the current obesity epidemic in the United States. Residents of sprawling neighborhoods are more overweight (have a higher body-mass index), have greater rates of obesity, have elevated hypertension rates, and do not walk as much as their counterparts from less spread-out neighborhoods.[58] These effects persist when models control for the effects of demographics and health behaviors. Sprawl even hurts the pocketbooks of residents in these neighborhoods, who have more private motor vehicles and spend a larger part of their income (greater than 15%) on transportation than residents in more compact communities.[59] Moreover, sprawl eats into people's free time, as commutes become progressively longer. Between 1960 and 1990, the percentage of people working outside the county in which they resided doubled.[60]

The effects of sprawl are not limited to those who choose to live in such areas. Unfortunately, sprawl harms the quality of life for residents outside of suburbia in ways beyond the issues of environmental damage, pollution, and human health. Sprawl contributes to urban decay, the perpetuation of racism, and declining sustainability in national economic growth. After the civil rights movements of the 1960s, governmentally sanctioned segregation was gradually replaced by sprawl-based segregation.[61] Sprawl occurred nearly twice as fast around metro areas with significant minority populations (a 3:1 ratio of land-to-population growth) as around metro areas that were primarily white (1.8:1), because white people were separating themselves from minorities by moving to suburbs. In a variety of metro areas, white flight to suburbia created rapid sprawl, declining urban tax bases, a diversion of tax money to infrastructures in the suburbs, and de facto segregation. This process created a segregation tax, where poor white homeowners paid 30%–50% more for equivalent housing in the suburbs, and blacks faced home values that were 30%–50% lower because whites would not enter the market for homes in the urban core.

Even if arguments appealing to social justice, environmental sustainability, and improved health don't provide sufficient motivation to address sprawl, self-preservation should. The public debate over peak oil (the point when oil production reaches its maximum and subsequently declines) has been quite controversial, with extreme positions positing unlikely scenarios, ranging from extreme optimists' expectations that oil is produced quickly underground by bacteria to extreme pessimists' predications of a cataclysmic end to modern civilization. A more moderate stance still paints a dire picture, particularly for regions plagued by sprawl. The Hirsch Report, the fundamental document regarding peak oil, suggested that the peak in production either occurred as early as 2005 or would occur by 2025, depending on whether data for the estimate are drawn from nominally independent researchers or Shell Oil.[62] The report also emphasized the fact that, among other issues, sprawl makes the United States more vulnerable to peak oil than most other nations. The report highlighted a few key findings relevant to the impacts of sprawl on humans' well-being:

- Although the timing is uncertain, peak oil will happen, and it will be abrupt, unlike previous transitions between energy sources.
- The impacts of peak oil on the gross domestic product will rest on how dependent national economies are on low-cost oil (the impact on U.S. economies could be measured on the trillion-dollar scale).
- Waiting until peak oil occurs will create 20 years of liquid-fuel shortfalls (with 20 years of hardship for people living in sprawl).

Mitigation strategies for dealing with the time when declining oil production intersects with exponentially increasing demand include initially making transportation more efficient, and subsequently squeezing oil out of tar sands and coal.[63] These approaches, however, ignore the possibility of an easier mitigation strategy: reverse sprawl, so people don't rely on long-distance travel in their automobiles in order to survive.[64]

Sprawl emerged in the late nineteenth century in the United States for several reasons. First, residents tended to view the supply of land as unlimited, so there was nothing inherently wrong with losing acreage to sprawl.[65] Second, property-ownership rights (including those associated with development) were protected by the U.S. Constitution. These property-rights views were so strong that restricting development on a piece of property was framed as a governmental taking of property, and some residents who personally opposed development nonetheless chose not to advocate for community planning, because they believed that they should protect property rights as a matter of principle.[66] Third, public distrust of government created a system in which development was highly codified. The rules removed public officials from decision making by treating development as a use-by-right, thereby exposing it to minimal public review. This tradition meant that developers creating far-flung subdivisions received limited public oversight. Conversely, the ordinances in zoning and development regulations lack language that would allow conservation subdivisions, mixed-use development, and higher density as a use-by-right, which means that developers must go through a lengthy and expensive rezoning or variance request before building subdivisions that do not create sprawl.[67] Despite conservation subdivisions costing up to 30% less to build and often having higher market values, the onerous process of obtaining permits for them has severely hindered their creation. Fourth, there was an unrivaled trust in the free market, paired with a poor understanding of it.

This trust in an often-misconstrued concept of free markets is expressed by claims that sprawl is simply an expression of personal preferences. The free-market argument suggests that consumers thoughtfully considered the costs and benefits of various possibilities and then chose to live in sprawl. Evidence does not support this supposition, however. First, research indicates that consumers consistently underestimate (by 60% or more) or ignore the costs of automotive transportation.[68] Second, if consumers tried to accurately assess these expenses, the amount would still be less than their true cost, reflecting the huge governmental subsidies for the infrastructures (roads, sewers, power grids, and oil exploration) serving sprawl that are paid for by commercial development and people living in urban centers.[69] Third, even if the costs of sprawl included current subsidies for infrastructures, they would not account for a legacy of governmental policies and international interventions designed to reduce energy costs.

The government policies and subsidies supporting sprawl in the United States explain why sprawl defines development in this country more than it does in Europe. There is compelling evidence that the greater or lesser presence of sprawl is due to political power exerted by the economic elite and producer groups in both regions.[70] Specifically, the different responses reflect energy politics. By the late 1800s, it was clear that the United States had far more fossil fuels available for energy-intensive industry and development patterns than did Europe. The United States was the leading oil producer until the 1950s and had massive supplies of coal and natural gas, while Europe had both limited reserves and fewer of the political and military resources required to extract a reliable energy supply located elsewhere. Accordingly, economic elites and producer groups in the United States promoted energy-intensive policies, while their counterparts in Europe promoted policies that reduced the risks of dependence on fossil-fuel supplies.

In the late 1800s, wealthy land speculators in areas around Los Angeles and Boston built and promoted streetcars as a way to increase property values for land they had purchased on the urban periphery, the first suburbs.[71] The streetcar lines were laid out to connect city centers with the developers' planned suburbs. Conversely, in Europe trolleys were designed to provide economical transportation conveying people into urban centers. Further, by the 1920s, the United States was the world's leading manufacturer of consumer durables, notably automobiles, and single-family houses—more than apartments or the size of the population—created a demand for more consumer durables. The 1926 Supreme Court case *Village of Euclid v. Amber Realty Co.* established zoning as being constitutional and set the precedent for conventional zoning in the United States. That decision allowed Euclid to prevent industrial land uses from mixing with residential ones. This pattern of zoning tended to quarantine uses from each other, and promote low density as a solution to congestion and pollution.[72]

Franklin Delano Roosevelt's response to the Great Depression included massive expenditures on expanding roads, bridges, and water and sewer systems from urban areas into the countryside. The United States developed programs to guarantee home loans, which made escaping mortgages by declaring bankruptcy far easier than in Europe. Thus one response to the Great Depression effectively became a concerted government effort to promote sprawl.[73] The U.S. pro-sprawl policies piled up even faster after World War II. Returning veterans received cheap and easy credit for home purchases, and benefited from tax policies and deductions that promoted home purchases.

Developers prepared for sprawl by purchasing and subdividing huge tracts of land adjacent to metro areas.[74] The 1945 Federal Housing Act provided federal support for installing water and sewer systems in expanding communities. The Interstate and Defense Highway Act of 1956 created an interstate highway network that funneled commuting suburbanites to urban jobs.

Gasoline taxes in the United States started in the early 1900s, and they were used almost exclusively to maintain and expand the infrastructure needed for automotive travel. In contrast, policies in Europe entail far larger gasoline and energy taxes that are used for general revenue, not just highway development. The strategic approach taken by the United States has served economic elites (including certain producer groups) well, but it cannot do so sustainably. The fact that the United States has been running an annual trade deficit of half a trillion dollars to pay for oil imports during the longest and most severe recession since the Great Depression underscores a dangerous future for sprawl-based economies.[75] In the next chapter, we will delve more deeply into some of the cultural and political rationales for these policies.

Landscape Ecology and the Housing Bomb

Housing has innumerable impacts on society and the environment, but many of the most important ones arise through indirect means. The direct impacts should be obvious, including the loss of natural areas, the loss of agricultural lands, damage to natural resources, and energy use. The indirect impacts, which are just as important, arise from interactions among households, society, and the environment that are more difficult to see. Landscape ecology is a discipline that combines ecology (the study of interactions between living and non-living parts of natural systems) with an explicit focus on spatial patterns and processes.[76] As such, it is well suited to identifying and evaluating the indirect effects of housing. Increasing numbers of households require more housing units, which convert forests, grasslands, wetlands, agricultural lands, and other types of land into residential areas. Land conversion and its associated activities destroy biodiversity, including plants, wildlife, and habitat for wildlife.[77] Another complex chain of impacts is started by the construction materials needed for homes. Production of the wood, concrete, steel, glass, and energy needed for housing construction and maintenance causes cascading effects, such as the fragmentation and destruction of forests and wildlife habitat far away from residential areas.[78] Household products in China made of tropical wood, for example, can affect forests thousands of miles away.[79]

Many of these complex relationships involve thresholds. For instance, housing is an impetus for the construction of roads. Initial increases in the numbers and types of roads have proportional impacts on wildlife and aquatic species, but once a critical threshold is reached, the road network will isolate plants and animals in small fragmented patches of the landscape.[80] Residential development within riparian habitats has strong negative impacts on bird communities, and, at high densities, can nearly wipe out avian communities.[81] On the one hand, riparian areas are essential to many kinds of breeding birds, and both these lands and the species that require them may be more sensitive to the pressures of development than other areas. On the other hand, some species that can eat a broad array of foods, and some that prey on bird nests, tend to increase with residential development. Although clustered development decreases habitat loss (in this instance, the rates of habitat loss are lower than the rates of housing growth), homes are often clustered in the most sensitive ecosystems (e.g., lakeshores and riparian areas), which contain critical habitat for many wildlife species.[82]

Scale often dictates the impacts of housing. At small scales, housing can serve either as barriers against or as corridors for the movement of wildlife. House walls may limit the spread of and colonization by insects, but plants along the pavement may act as dispersal corridors.[83] At larger scales, housing and the roads connecting houses to jobs, retail establishments, and recreation typically destroy dispersal corridors.[84]

Housing is a significant contributor to the invasion of plants from other regions. Householders introduce exotic (non-native) plants for landscaping and create habitat conditions suitable for these introduced species, which then may spread unchecked. For example, rural housing patterns explain the distribution of invasive non-native plants in temperate forests of the midwestern United States.[85] Housing variables (e.g., the number of houses in a one-kilometer [2.2-mile] buffer around each plot) showed the strongest association with the richness (abundance) of non-native invasive plants (such as buckthorn and honeysuckle) in this region.

Conclusion

The demographic transition in household size discussed above suggests that even if population growth stopped completely in 2010, the housing bomb would still include an explosive growth of new homes. If the global number of houses per capita stabilizes at 0.4 (2.5 people per household), as it has in developed countries, the change in household size would require 1 billion new homes for the current number of people, in addition to needing more homes per capita to accommodate any future population growth. Moreover, this may be an underestimate, given recent research highlighting a trend toward Americans choosing to live alone. In the United States, the number of one-person households grew from 13% in 1960 to 27% in 2010.[86] Unless new policies and economic incentives can be devised to reduce the demands for land, energy, wood, steel, minerals, and other resources normally associated with housing, the impending housing bomb will cripple both global economies and the natural ecosystems that supply the resources needed for human survival. Chapter 2 will describe why the needed changes have been hard to make, and the remainder of the book suggests ways in which those changes can be accomplished.

Perhaps the best way to end our discussion of household dynamics is to review promising adjustments that are already starting to happen. The Great Recession forced people to consider the costs of many housing decisions, and the unrelenting demand for oil has pushed prices inexorably upward, despite economic slowdowns. In this context, where the true costs of resource-intensive housing are more difficult to ignore, the values of homes in walkable neighborhoods have begun to outpace those in sprawling suburbia.[87] This shift reverses a historical pattern, where suburban homes accessible only by private motor vehicles cost more per square foot than other types of housing. Notably, this new trend exists across a wide range of neighborhood types, including exclusive subdivisions.[88] Total U.S. housing values dropped between 2008 and 2011, but sprawl-based suburbs experienced huge price decreases, while homes in close-in neighborhoods maintained their resale values or even saw price increases. Such trends are likely to accelerate as baby boomers dump their detached houses situated on large lots into the market at the same time that interest in this type of dwelling is being supplanted by demands for multiunit housing and small-lot homes.[89] These results suggest that a preference for mixed-use, compact, walkable neighborhoods is emerging and growing, despite policy and taxation legacies firmly supporting sprawl. Although this trend bodes well for urban renewal and the economic sustainability of future development in the face of peak oil, it does not necessarily address inequity issues associated with income and ethnicity, since it appears that gentrification (the purchase and renovation of decaying urban property by middle-class or affluent individuals) is affecting the most walkable urban neighborhoods.[90]

CHAPTER TWO

How Home Ownership Both Emancipates and Enslaves Us

In the twenty-first century, owning a single-family home has come to represent much more than the security and safety afforded by basic shelter. In many cases, it is seen as the visible demonstration that its owner has achieved some semblance of the good life presumably sought by all human beings. In the Introduction we discussed the phenomenon of house addiction, and argued that this addiction has led us on a self-destructive binge that could blow up in our faces. In chapter 1, we sketched a broad history of household dynamics and described some demographic drivers of the housing bomb. In this chapter, we focus on the social construction of relations between personhood and home. We ask how home ownership became so closely identified with personal happiness and social status, how owning a single-family home became synonymous with patriotism, and how these attitudes and beliefs influence household dynamics today.

It seems relatively clear that home ownership conveys benefits in the form of transaction savings, because once one has purchased a home, negotiations with a landlord are eliminated. In many nations, legislation and other rulings enhance the benefits of home ownership by providing legal means for homeowners to extract funds from society, including from those who are not homeowners. It also may contribute to efficiency, as homeowners accumulate local knowledge through longer tenure in a single location. But somehow, home ownership has developed a cultural significance that extends far beyond what would be suggested by these advantages.

Socioeconomic Context from the Twentieth Century

From today's perspective, it seems obvious that single-family home ownership is universally valued; it conveys status, security, and freedom to homeowners. When we use the term home ownership, we are referring to owner-occupied residential real estate, not structures purchased for other purposes. And, at least in the United States and Britain, the homes we are referring to are primarily single-family dwellings.

People have been congregating into various forms of municipalities for a very long time, at least since the era when humans figured out how to organize themselves in ways that enabled them to stockpile concentrations of surplus products.[1] The values and beliefs that undergird today's household dynamics, however, are much more contemporary. Although most analyses of the ideology supporting current home-ownership patterns have been based on British and other Anglo-Saxon cases (including the United States), the relevance extends far beyond them. These home-ownership patterns, although largely developed in the West, have been wholeheartedly exported to other nations and regions, such as Hong Kong, India, Singapore, South Korea, and Taiwan.[2]

Humans need some sort of shelter, and ownership of a single-family home is widely thought of as the preferred way of obtaining that shelter. When we were writing this book, the National Association of Realtors (NAR) was airing the following advertisement on radio and television throughout the United States:

> When you're a kid you don't know much about housing markets. You don't know that home ownership builds communities. And you certainly don't know that owning a home contributes to higher self-esteem and better test scores. You just know that home is where you belong. It's where you play, grow, and learn. The National Association of Realtors wants you to know that home ownership matters: to our families, our neighborhoods, and our country.

Audience members were encouraged to go to the NAR website to learn more about home ownership (www.realtor.org/videos/public-advocacy-campaign-moving-pictures/). On the website, we found a news release dated June 27, 2012. The text reassured readers that the NAR understands that "there's a reason home ownership is called the American Dream. . . . Realtors remain committed to keeping the dream of home ownership alive for generations of Americans" (www.realtor.org/news-releases/2012/06/home-is-where-the-history-is/).

Many in this country live the American dream of home ownership, but the British rate of home ownership slightly surpasses that of the United States. Margaret Thatcher's conservative variation on the idea of a property-owning democracy increased the proportion of British homeowners while contributing to increased disparity in both income and wealth.[3] The Housing Act of 1980 provided a powerful policy arm to Thatcher's belief in privatization.[4] It gave tenants living in social housing (rental housing owned and managed by the state and/or a nonprofit organization) the right to buy their residences, usually at a discount from temporarily inflated market rates. Proponents proclaimed that the Act enabled working-class British citizens to cross that stark divide between those who have capital assets and those who do not. At the same time, opponents argued that it would dispossess those who were the most needy.[5] Either way, the 1980 home-ownership rate of 55% grew to 67% by the time Thatcher left office in 1990, and later increased to beyond 70%. Although many of those new homeowners have now become homeless, due to the recent financial bust, their loss does not change the fact that, when given the opportunity to purchase their homes, residents of social housing did so in droves, despite outrageously inflated prices.[6] The perceived advantages of home ownership have led to increased levels of owner occupation in several nations. Although Anglo-Saxon countries have promoted home ownership more strongly than those in continental Europe, various modes of supporting home ownership are now practiced in most developed nations.[7]

Taking the United States as an example, historical accounts indicate that, although a preference for home ownership over renting has existed throughout U.S. history, the belief that home ownership is both a basic right of and a responsibility for all citizens was just reaching maturity in the years following World War II, and it required considerable effort on the part of both the public and private sectors to reach that point.[8] Although financing options were limited in the late 1800s and early 1900s, "it was not financial ability that kept old-stock Americans from buying homes; buying simply was not that important."[9] Further, home ownership conflicted with Americans' desires for mobility. Persuading Americans that they wanted to own their own homes required massive campaigns carried out by a partnership between the public and private sectors. We briefly describe the main characteristics of these campaigns here, and then suggest reasons why the nation-state poured resources into them. (Table 2.1 offers a summary of especially significant U.S. agencies and government-sponsored private-sector institutions associated with home ownership).

The federal government's foray into housing began in earnest in 1917, when the U.S. Housing Corporation started working with private developers to build "affordable housing" for workers who were contributing to the war effort. Federal campaigns, working in partnership with national and local

Table 2.1 Important U.S. housing programs during the twentieth century

Acronym/Abbreviation	Name	Established/Abolished
FHL Banks	Federal Home Loan Banks	1932
FHA	Federal Housing Administration	1934
NHA	National Housing Act of 1934	1934
USHA	United States Housing Authority	1937/1947 abolished and consolidated within the Housing and Home Finance Agency, then reconstructed under HUD
Fannie Mae (now a government-sponsored enterprise [GSE])	Federal National Mortgage Association (insures mortgages)	1938/1968 became a private GSE
HUD	Department of Housing and Urban Development	1965
Ginnie Mae	Government National Mortgage Association (channels global capital into the U.S. mortgage market)	1968
EHFA	Emergency Home Finance Act of 1970	1970
Freddie Mac (also a GSE)	Federal Home Loan Mortgage Corporation (expands the secondary mortgage market)	1970 (as a GSE)

real estate organizations, publicly associated owners of single-family homes with thrift, moral character, and citizenship, and homeowners were lauded as patriots. At the same time, the rhetoric of these campaigns associated renters and those living in multifamily dwellings with negative concepts, such as communism and radicalism.[10]

Campaign publications portrayed both multifamily housing and tenancy as undesirable, stating that "in the minds of many it threatens American institutions."[11] At the same time, real estate agents began using the presence of homeowners to signify a neighborhood's superior quality, such as in this advertisement in the *Atlanta Journal*: "Davis Street is settled with a good class of people, most of whom own their own homes."[12] As Secretary of Commerce, Herbert Hoover opined that home ownership "may change the very physical, mental, and moral fiber of one's own children."[13] Paul Murphy, head of the U.S. Department of Labor's Own Your Own Home campaign, proclaimed that "man must provide for his family by building a home, else he robs his patriotism of practicability."[14] Pronouncements such as these were inserted into pamphlets given out at women's clubs, work sites, and public schools.

Americans apparently took a considerable amount of persuading, and a publicity agent within the government proclaimed that "[dissemination through these venues] is exactly the thing needed to shove the whole housing and better homes ideas of the Department over."[15]

Prior to these campaigns, the National Association of Real Estate Boards (NAREB) had not focused on selling single-family dwellings.[16] As a signal of its new emphasis, NAREB wrote the following to its members:

> Realtors: You have a definite duty. Aside from your natural wish to dispose of property on your lists, you have a patriotic call. The nation looks to you for education! Put a shoulder to the wheel! Get into the human side of present conditions. Teach the nation that there is one duty [ownership of a single-family dwelling] none should shirk.[17]

NAREB accepted the responsibility to teach Americans that every male citizen needed his own single-family home to provide the ideal environment for family life, especially child rearing. Owning such a home contributed to thrift and independence. The desire to own one's (detached) home was described as natural, which may be why apartment living was cast as detrimental to both personal and family life. Further, home ownership conferred status rewards, because homeowners were proclaimed as bulwarks of the community and the American way of life. Apartment dwellers, on the other hand, were described as more likely to be socialists, criminals, and therefore contributors to social disorder in the neighborhood.

Robert Lands notes that "by imbuing home ownership with particular meanings, federal and private interests were able to entice Americans into adopting practices and frameworks that served a number of national interests, including the stabilization of land markets and the adoption of specific political frameworks."[18] Slogans such as "A man's home is his castle" and "Home is where the heart is" gained enormous popularity, and they are still in vogue today.

Even with these enticements, the state of home ownership in American society remained radically different from what we know it as in the twenty-first century. Before the National Housing Act (NHA) was passed in 1934, the banking system required prospective homeowners to produce down payments equal to or larger than the size of the loan, which was beyond the financial means of many Americans.[19] Thus, despite campaigns glorifying home ownership as the material embodiment of the American dream, the ownership rate hovered under 40% until institutional changes loosened the market further.

President Franklin Delano Roosevelt's programmatic response to the Great Depression included both the massive development of various infrastructures and institutional changes to encourage citizens to purchase homes. Housing, which remained the biggest untapped loan market in the United States, was the target of the NHA. The NHA was expected to stimulate the severely depressed economy by awakening this sleeping giant. The Act included the establishment of the Federal Housing Administration (FHA), to supply the needed institutional boost. This governmental agency would provide insurance to private banks, especially those that made loans to low-income families who had previously been unable to enter the market. The goals of the FHA were to improve housing standards and conditions, provide an adequate home-financing system, and stabilize the market, all through insuring mortgages. Insurance supplied by the FHA provided banks with a sense of safety, encouraging them to make loans to the low-income borrowers they had previously avoided. Since the new mortgages required lower down payments and offered loan packages spread out over a longer period, borrowers could qualify more easily. Not surprisingly, home ownership rose from 41% in 1941 to 57% in 1957.[20]

These home-owning patriots streamed out of urban centers and into the suburbs, and they did so in a highly racialized pattern. Despite the rhetoric, the new mortgage industry did not create a society where all residents (or even all citizens) could own their own homes. In fact, the transformation may have solidified the boundaries between homeowners and renters, along with the social status associated with each.

At roughly the same time as institutions such as Fannie Mae and the Veterans Administration were opening up the mortgage industry to previously ineligible borrowers, another governmental agency was producing housing for those who still remained outside the market. The U.S. Housing Authority (USHA), created in 1937, focused on providing decent housing, even for those who still could neither pull together the required money for a down payment, nor qualify for the liberalized loans. To do so, the USHA loaned money to municipalities. These municipalities, in turn, were required to tear down slums and replace them with public housing that would be available to the poor.

Concerns that the widespread availability of this public housing might interfere with burgeoning real estate development led to policy decisions that severely limited tenancy, restricting it to the very poor. Cultural traditions combined with political power to ensure that public-housing projects were vigorously segregated along racial lines. When the Supreme Court began striking down the constitutionality of segregation with decisions such as *Shelley v. Kraemer* (1948; 334 U.S. 1) and *Brown v. Board of Education* (1954; 347 U.S. 483), it became clear that racially segregated public housing could not be maintained. After 1954, blacks began moving into public-housing communities that had previously been reserved for whites. Many of the white tenants, their minds still teeming with racist fears, migrated to the suburbs as soon as they could put together a down payment.[21] Thus, at the same time that tenancy in public housing became increasingly identified with impoverished inner cities, home ownership became increasingly linked to America's prosperous suburbs.

The U.S. citizens who remained behind in the inner city, still unable to purchase a home, could be easily identified by the color of their skin. Despite decisions such as *Brown v. Board of Education*, segregation in the United States did not disappear overnight. The quickly constructed subdivisions that provided affordable new homes were energetically segregated, with full support from residents, realtors, developers, and lenders. If black citizens actually looked for a home in these locations, they would be told that nothing was available.

Both public and private policy colluded to maintain segregation between white citizens who could buy their own homes, and black citizens who could not. The FHA's *Underwriting Manual*, published in 1935, mandated that a neighborhood have "protection against adverse influences," noting that "important among adverse influences are the following: infiltration of inharmonious racial or nationality groups; infiltration of business or commercial uses of properties; the presence of smoke, odors, fog, heavily trafficked streets, and railroads."[22] The private sector was squarely on board. As late as 1950, the NAREB Code of Ethics stated:

> A Realtor should never be instrumental in introducing into a neighborhood a character of property or occupancy, members of any race or nationality, or any individual whose presence will clearly be detrimental to property values in the neighborhood.[23]

The FHA contributed further to congregating blacks in public housing projects by redlining (marking a red line around) minority neighborhoods as risky ventures that mortgage investors should avoid.[24] Although the practice of redlining was made illegal by the Fair Housing Act of 1968, it has had long-lasting effects on black and other minority communities. Contemporary inequities in private-property ownership by minorities can be traced directly to FHA regulations during the New Deal era. As white middle-class homeowners flocked to the suburbs, owning houses with federally insured mortgages, blacks moved into public housing that white tenants had previously occupied. Because minority neighborhoods were designated as blighted, mortgages were less easily available for blacks who wanted to purchase homes there. Without the financial incentives available for suburban development, property values for homes in inner-city minority neighborhoods plummeted. Moreover, although the new suburbs continued to draw services from the urban core, the urban tax base all but disappeared.

Home Ownership in the Twenty-First Century

Almost a century after politicians and property developers collaborated to liberalize mortgage standards as one means of reinvigorating the ailing U.S. economy, home ownership has achieved what Stephanie Stern calls "an exalted status and privileged position in American property law."[25] She cites bankruptcy protections, property-tax relief, and foreclosure reform as examples of what she terms "residential protectionism."[26]

Home Ownership as a Basic Human Right

Although it imposes tremendous costs on society in general, as well as on people who rent their homes and on apartment dwellers, the concept of single-family home ownership is protected from critical evaluation by its "stature of moral right."[27] Scholarship regarding property issues relies heavily on Margaret Radin's largely unsubstantiated theory that a special class of property, including homes, is rooted in the constitution of the self, so that home ownership becomes a basic human right, on a par with life and liberty.[28] Thus what might otherwise be seen as policies that encourage irresponsible behaviors— such as rent seeking, boosterism, and land speculation—have been promoted as morally unimpeachable. Rent seeking refers to a situation where an individual or organization obtains an economic gain from others without any reciprocal benefits coming back to society through the creation of wealth. We are using the term boosterism to describe promotional activities often associated with extravagant predictions and inflated prices for real estate. Land speculation often results from successful boosterism, with prospective homeowners purchasing risky property investments that offer the possibility of large profits, but also include a higher-than-average probability of devastating loss.

Before leaving the question of whether home ownership is a sacred human right, let us engage in a thought experiment. Imagine that private-property policies were truly motivated by the belief that one's home is an extension of one's self, a psychological prerequisite for mental health. In that case, there would be no reason to privilege a traditional single-family residence over other sorts of homes. Since each person is a unique individual, a person's home could be expected to come in any imaginable shape, color, and size to fit the person for whom it is an extension. When someone wants to make their home in a school bus, a railroad car, or a truck bed, however, property law not only fails to provide protections for such homes, it tends to prohibit them. If this imaginary scenario seems ridiculous, perhaps it is because home ownership comes closer to being a sign of social status than an extension of the self. But protecting someone's social status does not have quite the same moral authority as safeguarding someone's personhood.

If home ownership does, indeed, create an extension of the self, it makes sense to provide security for it, and society has done so, at tremendous cost. These protections encourage excessive investments in residential real estate, narrowing the options present in a society's investment portfolio.[29] State laws that cap property taxes on homes benefit longtime owners at the expense of both new buyers and renters.[30] Homeowner protections also increase the cost of credit for everyone else, a cost that is experienced most heavily by poor families.[31] Home protectionism makes community planning more difficult, and encourages sprawl (see chapter 6). Sprawl, in turn, exacerbates anthropogenic (related to or resulting from the influence of humans on nature) climate change by continually boosting demands for consumer goods and fuel for the motor vehicles that we rely on to make suburbia appealing.[32]

Given the tremendous costs of residential protectionism, it seems reasonable to expect significant empirical support for the psychological benefits of home ownership. Susan Stern found an astonishing lack of such support, however.[33] For example, although homeowners use their homes to display or advertise their identities, there is no empirical evidence that home ownership makes important contributions to a person's psychological health, or that a lack of home ownership threatens it. There is evidence, however, that dislocation occurs more often among renters than among homeowners, although it is not clear whether forced, rather than voluntary, dislocation is more frequent among renters. Homeowners may face forced dislocation due to mortgage foreclosures and the exercise of eminent domain. Renters may also encounter forced dislocation when these situations arise, as well as in instances when the needs and desires of the people who own the lodgings change. Dislocation, especially when forced, does contribute to short-term stress, but stress levels quickly return to the same levels people were experiencing prior to the dislocation.

The ownership of private property confers both comfort and status on people, especially those living in capitalist societies.[34] But there is a striking lack of evidence to support the emphasis on residential real estate as a special case of property. As Stern argues, "The empirical evidence indicates that the psychological and social importance of the home has been vastly overstated. The psychological value attributed to the home has masked rent seeking as moral conviction and greased the wheels of the residential protectionism machine."[35]

Home Ownership as Social Capital

Home ownership, by increasing social capital, has also been touted as a strong contributor to closely knit communities. Social capital, in this case, refers to benefits (often economic) derived from cooperation between tightly knit groups of people.[36] Theoretically, communities with more volunteer organizations, more people meeting and greeting on the street, and residents who have lived there for a considerable time would have stronger and more valuable relationships than communities where residents are isolated and have short residency periods. Some property theorists have argued that protecting home ownership increases social capital by creating strong and meaningful social ties within the community.[37] Again, the empirical evidence does not bear out the claims. It is true that neighborhoods dominated by owner-occupied houses are more stable than those dominated by renters, with owners moving somewhat less frequently. Rather than discovering links between residential stability and tightly knit communities, however, the empirical evidence indicates that neighborhoods dominated by homeowners display the same declining sociability as neighborhoods composed primarily of renters.[38] Further, as we will demonstrate in chapter 6, the dominant form of modern suburbia clearly harms social capital by forcing people into automobile-based transport and off of sidewalks and streets where they might interact with their neighbors.

Even in matters where there is a measurable difference between the behaviors of homeowners and renters, when models control for other variables, a rather confusing picture emerges. Although homeowners are more likely to vote and to join neighborhood volunteer organizations than renters, if home ownership is decoupled from tenure (the length of time spent living in a place), most of the difference disappears.[39] Length of tenure, by itself, appears to contribute significantly to social capital by increasing the number of weak ties between individual community members. Yes, home ownership is associated with a lesser degree of mobility (e.g., in changing residences), but any policy to further limit mobility will increase the level of residential investment in the local community.

The decision to limit where people live, however, is not without its down side. Luckily for the majority of people who remain relatively mobile through-out most of their adult lives, communities are bound together by a dense mat of weak ties, not by strong ties. Politically active communities rely primar-ily on weak ties, which are typically reestablished within 6–18 months of a move.[40] Limited mobility may lead to greater stability, but stability does not necessarily lead to culturally, economically, and politically vibrant communi-ties. We have all traveled through neighborhoods that seem to have grown old along with their owners, and although they may exude serenity, more often they just seem tired. Migration facilitates cultural, economic, and po-litical vitality, since migrants may introduce new norms, values, and social practices that might foster adaptability.[41] All communities face problems, and given the rapid pace of change in modern society, responses that worked ten years ago probably are not sufficient today. Mobility provides a relatively non-threatening reason to reshape social relationships and institutional structures that have become outmoded. Observing migrants as they learn the norms of their new community provides longtime residents with opportunities to rethink automatic behaviors. The reciprocal learning that emerges from the process of new residents becoming embedded in new social networks lends resiliency to both individual participants and the community.

Why Do We Protect Single-Family Home Ownership?

Although there are many reasons why we may choose to retain the complex network of special protections afforded to owners of single-family homes, we should not do so under the illusion that we are safeguarding people's per-sonal psychological health or conferring unique social benefits on the com-munity. The current approach to individual home ownership and housing development privileges a small group of elites and limits social adaptability.[42] Further, current household dynamics are devastating from an ecological per-spective. Concerted efforts to encourage home ownership clearly increase the human footprint on the environment, because multiunit buildings are far more common for rental markets than for ownership of individual units. Although national data on rentals is limited, the United States began collecting this information in 2012. Those data will confirm what we already know: rental housing is smaller in size and, on average, more dense than owner-occupied housing. The 2010 census data showed that single-family homes were nearly twice as large (2,500 square feet) as units in multifamily buildings (1,400 square feet). So what are the compelling economic and political rationales for the policies that have enabled the development and maintenance of this unsustainable system of household dynamics? What justifies providing greater protection for ownership of detached single-family homes than for ownership of other types of property?

From today's vantage point, it seems obvious that increasing the number of homeowners made financial sense to a nation struggling to dig its way out of an economic depression. Yet the promise of an immediate economic reward provides only a partial answer. In the United States, the strategy of using widespread ownership of small landholdings to alleviate political tension dates at least to Jeffersonian efforts to stabilize the young republic by spreading ownership of its productive land as broadly as possible. Jeffersonian agrarianism posited that because yeoman farmers lived on their own land, they were independent of society for life's necessities (food and shelter) and dependent on it for protection and laws.[43] Because they possessed property, they were bound to defend the political system that sustained their property ownership, and in defending that system they stabilized it in the face of attempts to alter existing political arrangements.

Although Jeffersonian agrarianism had portrayed urban dwellers in derogatory terms similar to those used later for apartment dwellers by the U.S. Department of Labor's Own Your Own Home campaign, the concept of property ownership as a stabilizing force easily transitioned from farm ownership to single-family home ownership. In the face of widespread fears of social unrest associated with domestic industrial struggles, urban riots, communism, and anarchy, ownership of a single-family home came to symbolize patriotism and loyalty to the United States. Robert Lands noted that in the 1920s, "federal initiatives publicly associated the homeowner with thrift, character, moral fiber, and citizenship. [Their rhetoric described homeowners] as patriots and family providers, the bulwark of the nation-state."[44] In 1923, Secretary of Commerce Herbert Hoover claimed that suburban home ownership provided "the foundation of a sound economic and social system and a guarantee that our society will continue to develop rationally."[45]

Suburbanization entailed a radical transformation in lifestyles, which has been especially useful for powerful economic and political elites, allowing them to maintain control over the populace. First, it created needs for new products, ranging from air conditioning to multiple personal motor vehicles for each household.[46] Second, the fact that middle-class home ownership has been dependent on heavy federal subsidies helped alter the political landscape and focused community organizing on the defense of property values. Third, as debt encumbered homeowners, suburbanites helped weaken the threat labor posed to management, because workers could not afford to go on strike; without a paycheck, they couldn't make the next month's mortgage payment.[47]

Using home ownership to maintain social order and legitimize capitalist socioeconomic relations is not unique to the United States. There is strong evidence that home ownership has been politically sponsored to stabilize civil society by offering citizens a stake in a property-owning system; and it appears to have worked.[48] Since the 1980s, private ownership of homes has been particularly effective in buttressing neoliberal economic policies. Even in China, a nation that has only recently emerged as a powerful capitalist economy, home ownership has contributed to the rapid expansion of a neoliberal ethos.[49] Amid Deng Xiaoping's economic reforms in the early 1980s, markets were significantly liberalized, an entrepreneurial class was assembled, and an economic elite was reconstituted. The members of the entrepreneurial class, while not among the elite, have become ardent supporters of the political regime that has granted them the right to purchase private property and to use the credit system. Because this class's new ownership status relies on absolute political stability, the growing inequality between prosperity and poverty across different regions is ignored in what David Harvey describes as "a radical means of accumulation by dispossession."[50]

The same approach is illustrated in a recent World Bank report on international development. It recommends that developing countries move toward the system of home ownership favored in the United States: "Occupant-owned housing, usually a household's largest single asset by far, is important in wealth creation, social security, and politics. People who own their house or who have secure tenure have a larger stake in their community and thus are more likely to lobby for less crime, stronger governance, and better local environmental conditions."[51] If the goal is to promote political stability, the best tactic is to make every citizen a homeowner.

Scholars, accepting the evidence that property ownership tends to stabilize the existing political regime, have turned to the question of who this stability serves, noting that while maintenance of an unjust social system may be ideal for the world's wealthiest people, others may find it less appealing.[52] This leads to questions of whether the urge to protect prevailing social structures led to the sanctity of home ownership, or whether it worked the other way around. From our perspective, it makes more sense to assume that prevailing social structures and the dominant ideology about home ownership have codetermined each other. Further, although we agree that it's important to know how we got to the point where home ownership is widely believed to be a fundamental human right, we are more concerned with understanding *how* current policies that have emerged from uneasy relationships between culture, economics, and politics drive contemporary household dynamics.

How Emancipation Becomes Enslavement

As we mentioned at the beginning of this chapter, home ownership conveys benefits to the individual homeowner, but those benefits do not necessarily translate into benefits for society. There is no question that current U.S. policy, both at state and national levels, redistributes wealth by extracting rent from non-homeowners and paying that money to homeowners.[53] The question is whether or not this redistribution is good for society. We should also note that the redistribution that occurs through residential protectionism is regressive. It privileges higher-income households (through a form of subsidy) at the expense of lower-income ones. The federal income-tax deduction for mortgage interest, a loss to the government of over $72 billion annually in revenue, provides larger subsidies (deductions) to higher-income homeowners, who generally have bigger houses, with more substantial mortgage payments. Similarly, renters pay property taxes indirectly through their rent (as taxes are usually a factor when landlords calculate rental fees), but landlords get to include these property taxes among the itemized deductions on their income tax forms.

Residential protectionism also slows society's ability to respond to shifting employment needs, frustrates land planning, and encourages people to sink all of their investment potential into residential real estate, which means that their retirement portfolios completely lack diversification. We might even argue that current policies regarding home ownership display blatant codependency and encourage irresponsible social behavior. For example, during the 1980s and 1990s, people in many regions of the United States purchased homes at grossly inflated prices, sinking all they had (and committing much that they did not have) into the irrationally optimistic hope of continued rapid price increases for housing.[54] When it became clear that the flood of foreclosures was not going to stop with just poor and blighted areas, the sub-prime mortgage crisis was declared. In response, the Housing and Economic Recovery Act of 2008 required lenders to forgive debts above 90% of the appraised value of owner-occupied residential real estate, and it then allowed the homeowners whose debts were forgiven to refinance with mortgages insured by the FHA. The federal government, supported by private mortgage lenders, stepped in to enable insolvent house addicts to continue their self-destructive behaviors. But, as with any codependency, the Act has a catch. When these homeowners eventually do sell their houses, they must pay the FHA at least 50% of any appreciation in the value of their homes. Moreover, if they sell within five years after refinancing, they have to pay the FHA an additional premium, calculated according to the number of years from the refinancing date to the sale.

Paul Krugman contends that, when removed from the elaborately woven safety net provided by the United States and several other nations, homes emerge as high-risk, undiversified assets that are far from the best investment for most people.[55] The claim that homes deserve protection because they are "an owner's largest asset" merely naturalizes (establishes as a common practice) an unhealthy cycle: a home becomes the owner's largest asset because of legislation that gives special privileges to residential property, and it then requires additional protection because it has now become the owner's largest (and sometimes only) asset.

Although non-homeowners bear the heaviest cost, homeowners also pay for the privilege of possessing a single-family dwelling. Social costs imposed on owners include limited mobility in response to changed personal circumstances and professional needs, and a disproportionate amount of their investments channeled into a single asset. Empirical evidence shows that, if home ownership does have an influence on people's basic psychological health, it causes psychological harm to low-income owners.[56] Further, home ownership actually has negative health effects for people who form the most vulnerable sector of the population, those who already suffer from poor health.[57] This is not to say that private-property ownership is not important. Some amount of private-property ownership *is* a high priority for people, but there is no evidence to support the special status given to single-family homes over other forms of private property.

Ironically, the independence supposedly accrued by owning a single-family home is heavily dependent on a complex network of property, bankruptcy, and tax laws.[58] Governmental subsidies perpetuate the myth of the independent homeowner at the same time that they undercut its believability. We are not arguing that subsidies are always a bad thing. There is nothing inherently wrong with using economic and other incentives to encourage behavior that results in either short- or long-term benefits for society. But it may not be a good idea to subsidize behaviors that do not provide significant benefits to society, and, in fact, impose significant costs.

The Subprime Mortgage Crisis

Homeowners, especially those who depended on mortgage liberalization to purchase their homes, are particularly vulnerable to economic cycles. In 2006, the rate of foreclosures within low-income areas of U.S. cities began to increase.[59] There was little outcry, however until mid-2007, when the foreclosure wave began to engulf white middle-class suburbs, especially in the South and Southwest. By the end of 2007, approximately 2 million people had lost their homes, and there was no sign that the problem was slowing down. By the end of 2008, in the midst of what the media had christened the subprime mortgage crisis, all of the major Wall Street investment banks had collapsed (or gone through forced mergers), and confidence in the global financial system plunged. The crisis didn't stop with the banks, but soon engulfed those institutions that had been set up to insure mortgage debt (Fannie Mae and Freddie Mac). By early 2009, the U.S. mortgage crisis had spread across multiple economic sectors, as well as the rest of the globe. The celebrated growth economies in South and East Asia, for example, began contracting at a frightening speed.[60]

Although the global economic meltdown involved much more than just household dynamics, that is where it began. And we see little evidence that people have learned from this crisis. A recent World Bank report on international development recommended that developing countries establish the very system that had so recently crashed in the United States, suggesting that "when a country's system is more developed and mature, the public sector can encourage a secondary mortgage market, develop financial innovations, and expand the securitization of mortgages."[61] Apparently home ownership, mortgage financing, and securitization will magically transform poor nations into economically prosperous and politically stable countries. The authors of the report either failed to notice, or did not care, that this system is currently leading to mass dispossession and the loss of assets for the most vulnerable populations in the United States, Britain, and other nations.

Despite the accepted definition of crises as events that pass quickly, the housing markets in the United States and numerous other nations have yet to rebound to pre-crisis levels. The subprime mortgage crisis may even pale in comparison to housing crises associated with baby boomers retiring and trying to sell their high-priced homes.[62] Dowell Myers and SungHo Ryu have noted that the 78 million baby boomers have driven up housing prices since the early 1970s, and the smaller and less economically well-off generations following them will not be able to buy these homes.[63] To make matters worse, the upcoming generation of home buyers is far less interested in huge homes on large lots than previous generations, so fewer will buy these homes even if they had the necessary capital.[64] Arthur Nelson estimated that in 2003 there were already 22 million more large, detached McMansions on the U.S. landscape than would be needed in 2025, and that the nation faced a shortage of 26 million units in multiunit housing and 30 million in detached small-lot developments.[65] This is good news for those advocating efforts to defuse the housing bomb, but bad news for the owners of enormous homes unless someone can devise creative uses for the gargantuan houses, such as dividing them into multiunit dwellings. The latter scenario may be far-fetched, however, since such homes are often far from employment and from public transportation needed by residents who would purchase units in multifamily housing.

The continued homage paid to the current system of home ownership makes little sense, unless the goal is simply to ensure that the wealthiest members of the population continue to distance themselves from the rest of the world. The economic elites of Britain, continental Europe, the United States, and numerous other developed nations compose considerably less than 1% of the world's human population. Yet these elites retain the ability to dominate decisions related to household dynamics and other important topics because they possess wealth and incomes that can be converted into other resources as needed. Nonetheless, crises repeatedly erupt. We see the globe's almost continual state of crisis as an indicator that change is possible. And what better place to start than with households?

What Current Household Dynamics Offer Society

Despite the fact that the individual emancipation provided by current household dynamics may have been oversold, perhaps continuing these patterns could be justified, because they contribute to society at large. Again, though, we come up against a whole suite of problems. A U.S. Supreme Court decision related to household dynamics illustrates how complicated this can be. In *Kelo v. New London* (2005; 545 U.S. 469), the Court ruled that the city of New London, Connecticut, was justified in using eminent domain to claim the land where 115 houses stood, in order to sell the property to private developers. In a 5-to-4 decision, the Court stated that New London was not violating the eminent domain portion of the Fifth Amendment (giving states the power to take private property for public use), because the plan promised to benefit the economic development of the entire community, rather than a specific party (individual developers were not named in the plan). Although the developers who purchased the properties benefited from the decision, there has been a strong backlash to the Court's ruling.

Despite the fact that we are not comfortable with the Court's decision to forcibly evict householders so private developers could take over their property, we find the passionate public reaction to be more important. Like public support for the liberalization of mortgage markets, public opposition to the Court's ruling united conservatives and liberals—a rare feat. Conservatives expressed ire that the ruling directly threatened private-property rights, while liberals voiced their anger because it granted wealthy corporations an indirect path to powers of eminent domain. But the public's emotional fervor goes far beyond these statements of political philosophy. The political fallout from this case has made planning more difficult, constrained economic development, and contributed to sprawl. As part of the public reaction to *Kelo v. New London*, nineteen states (including Connecticut, Ohio, South Carolina, and Utah) passed legislation restricting eminent domain to blighted areas.[66] One result of this legislation is that municipalities and other local governments are now, more than ever, driven to situate public projects farther away from economic centers, where land is cheaper and more easily acquired, even in cases where a public project may have a significant value for the community.

It also provides legal justification for municipalities to target poor neighborhoods when seeking a place to dump unwanted materials or undesirable development projects, since low-income areas are more likely to be designated as blighted than are wealthy neighborhoods. This concern may seem less relevant today, in a world where campaigns to reallocate assets to people currently living in poverty are loudly opposed,[67] and schemes to redistribute assets to those who already command the greatest share of them prompt only a raised eyebrow or, at most, a visit to the local Bank of America offices to sing

Christmas carols about mortgage foreclosures (www.youtube.com/watch?v=
Tns3sljBvxU&feature=topics/).

Although social justice may not currently elicit the considerable attention
that it has had in the past, any serious efforts to reconstruct housing dynamics will need to include it. Whether for purposes of social stability or for other
reasons, humans have effectively spread house addiction broadly throughout
the globe. The World Bank argues that everyone needs to own their home,[68] but
this widespread expectation simply intensifies the barriers between those who
are homeowners and those who aren't. In a world where mortgages are easy to
come by, people who do not own their homes stand out as especially problem-
atic; they may be seen as irresponsible, erratic, or even politically radical.

The present model of home ownership also fosters a military-compound
mentality, where homeowners' primary reason to join together is to protect
their most significant asset from invaders. Gated communities have become
the norm, noted Mike Davis in his analysis of Los Angeles in the twentieth
century: "As the walls have come down in Eastern Europe, they are being
erected all over [western cities]."[69] In a world where urban apartheid is fast
replacing racial apartheid, those relegated to the margins cannot be expected
to remain there forever.[70] And if we extend our attention beyond cities such
as New York, Baltimore, and London to include others like Beijing, Mumbai,
Cairo, and São Paulo, the likelihood of sustaining current household dynamics seems even more absurd.

Conclusion

Despite the fact that today's dizzying pace of development may seem frightening, it also offers the potential for productive change. There is no question that humans need shelter to survive, and current housing dynamics have
grown out of this basic need. The human footprint on Earth has expanded
dramatically over the past several centuries. Our efforts to house ourselves
have proceeded in fits and starts, and they have taken different directions,
depending on all manner of temporal and spatial constraints. Perhaps the
realization that we could simultaneously improve our own quality of life and
the environment in which we live will encourage us to experiment with some
alternative ways to shape our footprint on this planet. The appropriateness
of those designs will vary, however, depending on location and community
preferences.

Our success depends on our determination to make maximum use of what David Harvey called humanity's "basic repertoire" for action:

1. Competition (between people who have different interests and different positions in terms of their political power)
2. Cooperation (between people who participate in voluntary social organization and institutional arrangements)
3. Adaptation (when people respond to changes in their environment)
4. Spatial orderings (when people adapt by migrating from one place to another)
5. Temporal orderings (when people adapt by attempting to change the way processes cycle through time)
6. Transformation (when people attempt to fundamentally modify the world to suit their needs or desires)[71]

Harvey presents this repertoire as a summary of the human abilities urban residents should rely on when staking their claim to the cities where they live. Competition refers to the rough-and-tumble political conflicts that many may shy away from. Competition needs to be tempered by cooperation, however, if a society is to prosper. We might think of residents in a neighborhood engaging in cooperative competition as they attempt to adapt to various economic or environmental stresses, ranging from a loss of employment to water rationing. If they cannot sufficiently adapt or diversify in their current location, people may be forced to reorganize their space by migrating to another community where they can find a job. Or they may try to negotiate a delay in the due date of their next mortgage payment or their rent. If individuals are not able to cope by personal adaptations, they may even attempt the more radical stance of transforming the world around them. They could join an organization of disaffected homeowners who are refusing to leave their homes (despite being unable to meet monthly mortgage payments). If drought is the problem, they could campaign to change a neighborhood prohibition against xeriscaping. Although specific practices would vary dramatically across different communities, the above repertoire suggests a framework for the conscious integration of humans' capabilities. It holds great promise for incrementally defusing the housing bomb—in the United States of America, the People's Republic of China, and many other locations across the globe.

CHAPTER THREE

"Housaholism" in the Greater Yellowstone Ecosystem

A feature story in *High Country News* began

> My name is Susan; I live on a ranchette. In the growth-pained West, this is
> as serious a confession as alcoholism or cruelty to animals. A year and a half
> ago, I picked up my local newspaper in Bozeman, Mont., and there under the
> headline TRACKING SPRAWL was an aerial photo of the Bridger Mountain
> foothills that included my place.[1]

Susan is one of the Greater Yellowstone Ecosystem's (GYE) many residents
who fell in love with the land in part by living on it. The remainder of Susan's
confession of house addiction reflects the household dilemma many environmentalists face in the GYE. Her 20-acre ranchette made her among the worst
contributors to sprawl, but living on that 20 acres tied her to that land and its
wildlife and plants. Her love for the place made her a supporter of clustered
houses; of strict covenants regarding water use, pets, and landscaping; and of
the Greater Yellowstone Coalition (an environmental organization dedicated
to protecting the wildlife, water, and lands of the GYE). Ultimately, Susan
found it heartbreaking to contemplate the idea of protecting the ecosystem by
removing humans from it versus living on it. She could not advocate ejecting
people from the land, but she knew the constant sprawl of new housing was
threatening the wildlife, waters, and landscape that she loved.

Yellowstone National Park highlights this dilemma. The park is touted
as the world's first, but nature parks had existed before 1872, when Yellowstone was founded. Rather, Yellowstone started the North American model
of national parks, where people were encouraged to visit, but were banned
from living on the land. Most maps of the GYE reflect the tradition of focusing conservation efforts on areas where people are prohibited from living by
creating cookie-cutter holes and divots out of the GYE that represent residen-

tial and agricultural areas—including Driggs, Idaho; Big Sky and Livingston, Montana; and Dubois, Pinedale, and Jackson, Wyoming. The maps clearly indicate that when you move into human habitat, you move out of the GYE.

Yellowstone National Park and several wilderness areas in the region are based on the premise of removing a permanent human presence to protect nature. They also serve as a constant reminder of the dissonance between the desire to protect the GYE from humans and the desire to build a house on it. The choice between living on the land and protecting it was not one Susan was willing to make, so she hoped foregoing her allotment of "2.5 children" and making major contributions to Planned Parenthood would provide karmic mitigation for her 20-acre ranchette. Unfortunately, human fertility in the region has little impact on population growth or sprawl. Until the 1970s, the GYE's highly fertile residents saw their children leave for places with jobs, and the population remained fairly stable. Both population growth and sprawl boomed after 1970, however, and these phenomena were driven by immigrating nature lovers. The GYE's livestock producers and farmers weren't contributing to population growth by having children, they were contributing to population growth by selling subdivided land to immigrants (newcomers) seeking natural amenities and their own piece of nature.[2] There are still plenty of people eager to carve out their own piece of the GYE as soon as they can afford to.

The piece of paradise this chapter addresses is the Teton Valley (fig. 3.1). Teton Valley forms a divot on the western edge of the GYE, and the valley was labeled the top private land-conservation priority for the entire GYE in 2002.[3] When Ed and Ellen Peterson moved to Teton Valley in 1950, they became what the locals there call "implants." Ed built his home quite literally on the banks of Trail Creek. It was the last house on Wyoming Highway 22 before the road crossed into a national forest and climbed over Teton Pass. The 900-square-foot house was built on the bank of a trout stream, before words like sprawl or riparian area were part of the local vocabulary.

Over the years Ed carved terraces out of the creek bank for a garden. His most cherished times of the day revolved around working in the garden and fly fishing in one of the three fishing holes situated on his 1.5-acre home site. He knew each hole in microscopic detail. The first was a small riffle behind a fallen log, the second was a deep spot in front of a beaver hole in the creek bank, and the third was a sharp bend in the creek where water gathered. Ed was concerned about his local stream sites, so he volunteered to measure sediment loads for nearby waterways and would sample water from them every week. Ed continued fishing and monitoring sediment loads long after

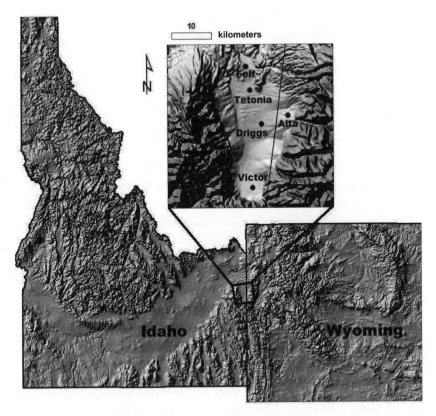

Figure 3.1 Teton Valley

he was legally blind. His work as a volunteer monitoring steam sedimentation eventually stopped after his memory of the roads faded and he started running his pickup truck into hillsides, but he never stopped his daily walk down the stream bank to his garden and his fishing holes. The last day he made that trip he felt unusually weak trying to climb the bank to his house, so he visited a doctor. He was diagnosed with terminal leukemia and died a week later. Ed's life was enriched by a profound sense of place rooted in his experience of living on the land, but he died before the wreckage left by the housing crisis in Teton Valley became apparent.

Household Dynamics in Teton Valley

When we arrived in Teton Valley to study household dynamics in 2004, Ed's house had progressed through an evolution, or devolution, of home uses: a home for a large family, a multigenerational household, a home for retirees, a home for a single person who was passionate about backcountry snowboarding, and finally a second home for people whose permanent residence was elsewhere. When the house was built, Ed was bucking a trend of reverse sprawl and depopulation. People were leaving the valley in search of employment opportunities. The few who could stay were moving, with their schools and churches, in a reverse-sprawl pattern: heading from far-flung regions of Teton Valley to the communities of Driggs, Victor, and Tetonia. The area's population had declined from a high of nearly 4,000 in 1920 to 2,600 in 1960, and it would continue to drop until 1970, when it bottomed out at 2,300. Teton Valley was settled by Mormon pioneers fleeing religious persecution in the 1880s precisely because "no one" knew about the valley. That relative anonymity, however, left the valley in an economic quagmire for decades.

Teton Valley flirted with discovery for years. When Jim Bridger first saw the valley in the early 1800s, he reportedly exclaimed: "This is the most beautiful valley in the world!"[4] For the first few decades of the 1900s, Victor, Idaho, was the terminus of the Union Pacific Railroad and served as a staging area for tourists traveling to Yellowstone National Park. The impact of the railroad slowly faded as automobiles replaced trains as a form of tourist transport. During the winter of 1936/1937, the Union Pacific Railroad considered the area for the Sun Valley ski resort (that eventually ended up near Ketchum, Idaho), but the close-knit Mormon community strongly opposed the idea. In the same period, Alma's Lodge emerged as a premier fly-fishing destination on the Teton River, and it was featured in *Sports Illustrated* in the 1960s. In 1964, after four decades of population decline, a group of local residents decided to sit down and hammer out a plan to provide local employment for their children. Jaydell Buxton, a participant in those meetings, described the situation:

> Back in 1964 . . . I don't know the exact figure . . . but there were only about 2,400 people in the entire Teton County. And if you didn't have a farm to go to or were a professional like a doctor or a pharmacist or something, you had to leave the valley. You couldn't stay because there was no work. There were a few tractor-implement dealers and such, but they were going pretty fast, because the new age of agriculture was coming upon us where new machinery was more expensive and farmers were having to run more land to pay for it. Everyone was having to leave and there was no future. So there was a group of people, I think almost all the valley participated, and we tried to think of some

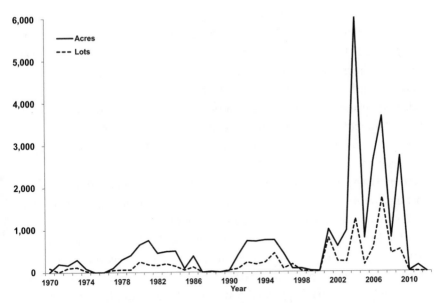

Figure 3.2 The number of new lots and acres platted for development in Teton County.

things to do to generate new industry. And we came up with quite a list and among those was the Grand Targhee Ski Resort. And of all of those things we had listed on there, that was the only one that came to *fruition* really . . . the only viable one we had. And once we started that, it really changed our valley. Almost immediately, things started to happen.

Grand Targhee Ski Resort was dreamed up, built, and paid for by a consortium of Teton Valley residents hoping to create a future for their children in the valley, and, as Jaydell said, things happened fast. The area's population decline reversed almost immediately, and Grand Targhee became the largest employer in Teton Valley. The people participating in the 1964 meeting had succeeded, but they got more than they bargained for. They set the community on a trajectory beyond the control of local residents. Initially, population growth was modest, with 20% increases in both the 1970s and 1980s. Then development exploded (fig. 3.2).

Media coverage, urban flight, telecommuting, excess capital from the dot-com and housing bubbles, and a demand for affordable housing from service-industry workers in nearby Jackson, Wyoming, converged to create a perfect storm of sprawl. In 1996, an article in *Men's Journal* listed Driggs as one of "America's coolest mountain towns" that "nobody's really known about—until now."[5] In 2002, *Men's Health* featured Driggs as the number-one "best place to live" out of 50 other contenders. The article cited Targhee's skiing as the "recreational linchpin" in the valley, and featured mountain biking and trout fishing.[6] That same year the movie *A River Runs Through It* was released, bringing world-wide attention to the lifestyle opportunities in the GYE. *National Geographic Adventure* magazine featured Driggs in 2001 as a hot spot for outdoor adventure access: "Friendly and unpretentious, Driggs is for people who love the mountains enough to live, work, and play in them."[7] *National Geographic* featured Teton Valley two years later and touted the local fishing, hunting, climbing, skiing, paragliding, ballooning, snowmobiling and skateboarding options as activities for outdoor enthusiasts.[8]

Rob Marin, who discovered the area during the decades of slow growth, saw the handwriting on the wall. In his 2002 article, boldly titled "Leave my town out of your 'Top 10,'" he asked, "How can our community avoid becoming a high-priced, badly planned playground for absentee homeowners who read too many magazines?"[9] Unfortunately, 2002 was far too late to be bringing up questions about avoiding an unplanned real estate boom in Teton Valley. The questions also came too late to avoid a housing crisis that damaged the economic, ecological, and cultural fabric of the community.

The media coverage was adding fuel to a fire that was already roaring. Jackson, Wyoming, has been "discovered" decades earlier, and that town had failed to address many key objectives for community planning (e.g., affordable housing for laborers, and protecting wildlife migration corridors). As a result, by the year 2000 more than 1,000 Jackson workers were living in Teton Valley and commuting over Teton Pass on a daily basis.[10] We found the steady stream of cars and trucks roaring by Ed's house, trying to gain speed before hitting the steep grades of the pass, so incongruous that it was comical—until we read the highway-safety statistics. This insane commute over a tortuous and often icy mountain pass wasn't just wasting time and gasoline, it was wasting lives. In terms of fatalities and serious injuries, the portion of Highway 22 crossing Teton Pass between Teton Valley and Jackson was one of the most dangerous routes in Idaho and Wyoming. Thanks in large part to the Teton Pass commute, Teton Valley was in one of only three counties in Idaho where the number of collisions increased instead of dropping in 2006.[11]

The Teton Pass commute also highlights another element of the household dilemma. Some environmentally oriented house addicts, including David Owen (a proponent of urban living),[12] have attempted to minimize the negative environmental impacts of their housing decisions by purchasing an old home in the countryside. The reasoning behind this was that buying an old home alleviated the need to build new ones. The logical problem with this rationale is that the previous owners do not disappear. They take the money, build a new home even farther from town, and start making a 70-mile commute over a huge mountain pass every day instead of a 50-mile commute. Similarly, the next person in line to live in the exurbs, who would have bought the old house, just switches to a new one.

Teton Valley was more than a bedroom community for Jackson, Wyoming. Newfound dot-com wealth and the employment flexibility associated with online technologies gave individuals the option to live and work in regions far removed from urban centers, and people were eager to leave.[13] During the 1990s, urban flight caused the greatest population increases in rural areas that had a high natural-amenity value for potential residents.[14] Outdoor recreational opportunities (notably skiing), an improved quality of life for raising families, or similar social factors motivated this migration, particularly to rural mountain areas.[15] Paul Starrs labeled those seeking a better life "urban refugees," who flee negative aspects of the urban experience.[16] A Teton Valley realtor we interviewed in 2004 said he moved to Teton Valley in 1994 "for the quality of life, the activities, the sports. I didn't have to worry about the kids getting run down and shot. I could walk down the street here and always see somebody I knew. I like small communities."

During the 1990s (according to the U.S. Census Bureau), Teton County was the fastest-growing county in Idaho, which was the fourth-fastest-growing U.S. state. The county's population blossomed from 3,439 to 5,999 residents (a 74% increase) between 1990 and 2000. Instead of people raising families, the newcomers were largely retirees, telecommuters, or other people seeking natural amenities. Hence the average household size declined, and the number of houses grew significantly faster (an 85% increase) than the population. Conservation developments and community planning largely failed to gain a foothold during that period, but land trusts exploded throughout the GYE.[17] Teton Valley was a forerunner in the land-trust movement, and the Teton Regional Land Trust (www.tetonlandtrust.org) was born in 1990. Ed's house sits beside one of the land trust's priority conservation lands (acreage that has yet to be acquired). The land trust protected almost 10,000 acres in Teton Valley, and it actually outpaced development with the amount of acres it protected until about 2004.

Ironically, the insecurity associated with the crash of tech stocks in 2000 and the 9/11 attacks jump-started real estate speculation in Teton Valley after three slow years in the late 1990s (fig. 3.2). In 2004, the year we started interviewing Teton Valley residents, 6,000 acres were platted for development, more than in the 1980s and 1990s combined. Until late 2007, residential home sales grew by 30% annually, and residential lot sales by 45%.[18] This explosive development served as a flashpoint for tensions between old-timers and implants. Accusations of hypocrisy among environmentalists and cronyism on the Planning and Zoning Commission were common during heated contests for county commissioners' seats. By 2005, the number of Teton Valley residents who had immigrated since the boom in 1990 formed the majority of Teton Valley residents, and in the next year's elections, commissioners who were willing to institute a building moratorium to buy time for community planning gained a majority.

Developers and investors saw the moratorium coming, and residential lot sales jumped by 145% in 2006. That same year, 2,000 lots were recorded. The 2007 building moratorium was short lived (March 27–April 30), and it generated community animosity instead of community planning. The Teton County Board of Commissioners approved a moratorium prohibiting land-use applications, rezone permits, and applications for subdivisions and planned-unit developments. The moratorium was justified on the premise that haphazard development posed "imminent peril . . . to the public health, safety, or welfare." On the evening of March 27, at an emergency public hearing, one of the two commissioners voting for the moratorium reported receiving personal threats after the vote, and the lone dissenting commissioner pleaded with the public not to physically harm the other commissioners or members of the public supporting the moratorium. A lawsuit seeking to reverse the moratorium was in the works before the moratorium was even passed.

On April 30, Idaho District Court Judge Jon J. Shindurling filed a decision that put an end to the emergency moratorium enacted on March 28, 2007. His decision maintained that Teton County's justifications for the moratorium (the county's lack of a garbage-disposal facility, a capital-improvements plan, a land-use map, and a planning staff) "show[ed] insufficient substantial and competent evidence to support the findings of imminent peril. . . . In fact it is debatable whether they constitute even a scintilla of evidence of such peril." He went on to argue that only something such as an imminent mudslide would justify the emergency moratorium. A housing bomb might have fit the "imminent mudslide" definition, but that wasn't considered.

Rescinding the moratorium allowed the pressure in the housing bubble to keep building until the Great Recession hit in 2008. By the end of 2008 there were 4,460 residential units in the Valley, and the vast majority (3,050) were built in unincorporated areas.[19] Platted lots exacerbated the trend toward sprawl, because there were 7,000 vacant lots platted in unincorporated areas of the valley, and only 7% of those vacant lots were within the expansion boundaries of incorporated areas. Residential home sales dropped by 40% during 2007–2008, and almost no sales were made in 2009.[20] Similar patterns occurred for residential lot sales, with a 10% drop in 2007, an 85% drop in 2008, and little activity in 2009. According to Alliance Title and Escrow Corporation (with a branch in Driggs), there were more than $104 million in foreclosures in Teton Valley in 2009, and more than $124 million in foreclosures in the first eight months of 2010. Nevertheless, in May of 2010 there were 28 subdivision applications still pending, for 1,796 units. When the housing bubble burst, there were more vacant lots (8,005 platted and 2,653 pending) than people (8,883) in Teton Valley.

The vacant lots left Teton Valley with "zombie subdivisions" scattered across the landscape. Even with the assumption that construction would recover to its 2009 levels, it would take 270 years to fill the subdivisions already platted. The timeline is even longer for unincorporated areas, with 340 years needed to fill those vacant subdivisions. If development jumped back to the previous above-average levels of 100 homes per year, it would still take 90 years to absorb the lots in unincorporated areas. Once the dust started settling, it became apparent that unplanned development had scarred the community. Newly paved roads ended in piles of gravel, with scarified lots stretching into the distance. Freshly poured concrete sidewalks were overgrown with weeds. Promised neighborhood parks were left partially completed. Economic, cultural, and environmental destruction had occurred across the valley.

Although the most obvious economic effects of the housing crisis in Teton Valley revolved around the halt of the thriving residential-construction industry and the catastrophic loss in personal wealth associated with foreclosures and declining home and real estate values, unplanned development has also locked in long-term economic woes for the community. If the vacant lots are built on, those homes less than one mile from county roads will require a total of $680,000 in county revenue to cover annual operations (e.g., filling potholes and repairing bridges) and capital costs (e.g., new buildings, roads, and police vehicles).[21] Moreover, building on the lots that are farther from county roads will generate huge annual deficits (1–3 miles: $8,188,205;

3–5 miles: $5,739,243; 5+ miles: $2,285,899). In sum, building on the platted lots in the valley would create an annual budget deficit of $17,427,220, which is astronomically high for an area the size of Teton Valley.

Projected costs associated with transportation and law enforcement cannot be addressed using the current tax structure in the county.[22] In the past, Teton County had relied on property taxes from vacant lots ($1,500 each annually in 2007) and a pyramid scheme where one-time fees for many development projects offset the costs of the relatively few homes requiring services. Since there were twice as many vacant lots as actual homes in Teton Valley in 2007, services were subsidized in part by the vacant lots. Unfortunately, it takes more than three vacant lots to offset the cost of just one occupied home. The tendency for homes to be built in more far-flung reaches of the valley exacerbated the budget problems, because taxes from more than nine vacant lots were required to offset the costs required to provide infrastructure for a home three to five miles from county roads. This shortfall was further complicated by the loss of vacant lots from the tax rolls, due to foreclosures and declining tax revenues through the devaluation of vacant lots. Subdivisions with names like "Shoshoni Plains" and "The Arbors" rapidly became delinquent on their taxes, forcing the county to foreclose on entire developments.

The economic devastation left by the housing crisis continued to expand as developers went bankrupt and walked away, leaving behind incredibly one-sided development-agreement contracts. During the housing rush, developers and their lawyers frequently wrote their own contracts and let the county sign them. In the permitting frenzy, it seemed as though some contracts were either never read or at least not carefully read. After the housing crash, people pulled out the contracts and noticed that many had no deadlines for when the developers were required to finish projects, and some committed the county to finishing the developments if the developers went bankrupt. To make matters worse, during the bubble everyone was gambling with land, and contractors would take lots in lieu of cash as payment for building roads, parks, and sewer systems. In these cases a contractor, paid in lots, had legal recourse to demand services for an entire subdivision from the county after the developer folded, even if only one home was actually built.

Second, the housing crisis had environmental consequences that put Teton Valley atop the list of the most-threatened sites in the GYE.[23] In their assessment of the GYE, Reed Noss and colleagues labeled the area as irreplaceable, because it had diverse habitats for most of the threatened species in the GYE, sensitive wetlands, migration corridors for mammals and birds, and essential low-elevation habitat (winter range) for ungulates, such as elk (*Cervus canadensis*) and mule deer (*Odocoileus hemionus*).[24] After polling conservation experts from the region and discovering high risks posed by development, they listed Teton Valley as vulnerable. The post-1990 development in the valley coincided with the first *E. coli* outbreaks there and with the near extirpation of native Yellowstone cutthroat trout (*Oncorhynchus clarkii bouvieri*) in the Teton River.[25] In the GYE, development threatens water quality, wetlands, migration corridors (75% of which were already blocked by 2004),[26] and winter habitat for mule deer and elk. The population of sandhill cranes (*Grus canadensis*) using Teton Valley to feed in preparation for a long migration to New Mexico fell by half as development, and the preparations for it, wiped out grain fields and wetlands the cranes used for foraging. Other species in need of serious conservation efforts that are threatened by development include breeding long-billed curlew (*Numenius americanus*), wintering trumpeter swans (*Cygnus buccinator*), and Columbian sharp-tailed grouse (*Tympanuchus phasianellus columbianus*).

These environmental impacts would be compounded if and when vacant lots are built on. Tripling the number of houses would cause irreparable damage to this irreplaceable component of the GYE. It also would triple the motor vehicle miles traveled, to 120,000 miles daily. If current trends of building farther from incorporated areas continue, the projected environmental impact may be underestimated. Presently, 6% of the homes that are more than five miles from county roads account for 20% of the vehicle miles traveled, and 40% of the homes that are less than one mile from county roads generate only 6% of the vehicle miles.[27]

Third, the housing crisis had major unintended consequences for the cultural fabric of Teton Valley. When *Men's Health* was touting Teton Valley as the best place to live, they made light of cultural tensions by suggesting that "if Driggs was a car, it'd be a tricked-out El Camino, tailgate smeared with the blood of last year's elk, hay scattered in the bed, a ski rack on the roof."[28] *National Geographic* gave slightly more gravity to the situation, describing the valley as "a community struggling to hold on to its way of life as 'newcomers' and new ideas flood into town like spring snowmelts."[29] The cultural tension was palpable as we started interviewing community members in 2004.

One old-timer who served on the planning and zoning board saw his role as protecting the rights of farmers whose land was "their 401(k)." He saw development restrictions as the hypocritical efforts of "implants" to steal retirement money from hard-working farmers. When he started describing the leader of the Valley Advocates for Responsible Development (VARD), fire shot from his eyes: "John Smith lives smack dab in the middle of what would be, by every one of his definitions, a wetland." VARD was the self-appointed watchdog for responsible zoning in the community, yet it had no employees or board members who were born in Teton Valley. During the first week of interviews we conducted, a middle-aged woman answered the door. After hearing about our study, she said, "I won't talk to you people" and slammed the door. A week later the same woman tracked us down and apologized profusely, saying that she should never have treated anyone that way. We were surprised by the apology, given our experience with interviewing elsewhere, and even more surprised by the ensuing explanation. The woman described a situation where immigrants saw the valley as a quaint recreation haven. When they moved in, they committed two unpardonable acts: (1) trying to steal old-timers' property rights by forcing development restrictions on the area, and (2) destroying the culture they considered so quaint in the first place. She felt that implants and academics like us were treating old-timers as if they were too ignorant and backward to recognize their own dilemma or to derive solutions for it.

There was some truth to her concerns about a lack of appreciation for local culture. We interviewed one immigrant in her home, a remodeled Mormon church that had been built by some of the valley's earliest settlers. Sitting in her remodeled church, the woman described how happy she was that "culture" was coming to Teton Valley in the form of yoga classes, a java shop, and a trendy restaurant. The president of the local arts council said he was surprised he ended up in the valley, because "it was such a culturally void place for many years." He added that "some of the locals have an amazing talent for quilting," but he worried that they would not "take it to the next level." Proponents of community planning often blamed old-timers, including the Mormons: "If there is a group of people that I'm the most disappointed in, it's the LDS [Latter-Day Saint, or Mormon] people who have been so quick to sell off the valley. On the one hand they claim family values and the quality of their social environment; at the same time they're diluting it just as fast as they can put their money in their pockets."

Previous studies in the Intermountain West have suggested that "culture clash" was overblown as an explanation for poor community planning and conflicting environmental attitudes.[30] So why did the housing crisis do so much damage in Teton Valley? Perhaps the failure to address community planning emerged from something more profound than cultural differences:

a dangerous cocktail of house addiction and property-rights beliefs. When we lived in Teton Valley during 2004, we saw a wicked combination of house addicts destroying what they loved, and property-rights fanatics refusing to deny anyone the freedom to develop anything. Were environmentalists suffering from house addiction and destroying ecologically sensitive areas of Teton Valley, and, if so, did they realize what they were doing? Were the community members who hated to see cultural and environmental change really fighting community planning in the name of property rights? By attempting to answer these questions, we were testing the premise behind the oldest joke told in towns struggling with sprawl from nature-seeking immigrants: "What's the difference between an environmentalist and a developer? The environmentalist already has his house in the mountains (or on the beach/ lake/river, or in the desert)."

House Addicts Destroying What They Love

We tested two hypotheses in our efforts to determine if house addicts were really destroying what they loved: (1) respondents with more environmentally oriented attitudes, and higher education levels, preferentially choose household locations in natural areas, and (2) the environmental impacts of household-location decisions are magnified by the smaller household size (fewer people per household) of people choosing to live in natural areas. The first hypothesis reflects the essence of the house addict's dilemma. Were people who loved nature, and were educated enough to know the impacts of their housing decisions, most likely to build homes in sensitive natural areas? The second hypothesis emerged logically from the first. The quotation at the beginning this chapter personifies the stereotypical, highly educated house addict who hopes a small family size can offset the impacts housing decisions have on the environment. If the stereotype holds, then environmentally oriented people would not only choose homes in natural areas, they would also create more homes per capita in those areas, by virtue of having fewer people in a shared home. Teton Valley was a great location to tackle these questions, because most people there were immigrants who had factored natural amenities into their decision to purchase a home, and minimal development restrictions meant that newcomers could build homes wherever they wanted.

We interviewed 416 Teton Valley residents in 2004.[31] Immigrants to the valley could choose to live in residential areas, agricultural areas, or natural areas. We mapped each respondent's home and classified it, based on the location the homeowner chose.[32] All households within city limits were categorized as residential. Natural land-cover designations were limited to wetlands and riparian zones (land along the banks of streams and rivers) on the valley floor, and forest or rangeland areas on hillsides bordering the valley. Homes surrounded by fields of crops were classed as being as within agricultural areas, unless they also occurred in a wetland or riparian zone. We determined homeowners' environmentalism with the new ecological paradigm (NEP) scale,[33] a series of questions designed to measure respondents' views on human overpopulation, on whether non-humans have rights, on how fragile nature is, on how likely future environmental disasters are, and on the degree to which people need to follow the laws of nature. Environmentalists (e.g., members of known environmental organizations) consistently score higher on the NEP than the general public or members of non-environmental organizations.[34]

Homes in existing residential areas needed very few new infrastructures (e.g., roads, sewer lines, and power lines), and caused minimal fragmentation of the natural land-cover. Homes in agriculture areas required road and power-line construction, and either an extension of current sewer lines or, more likely, the installation of septic systems; they also increased the number of vehicle miles driven by the homeowners. Finally, homes in areas with natural land-cover required all-new infrastructure construction, led to even more vehicle miles driven, replaced natural land-cover, and magnified their environmental damage by being either immediately adjacent to wetlands and streams, or directly destroying and otherwise fragmenting critical elk and mule deer winter range on low-elevation hillsides.[35]

Table 3.1 summarizes our household-location findings. Most (72%) of the people who bought houses in natural areas moved to Teton Valley specifically because they appreciated and desired the amenities nature offered. Conversely, less than half of the immigrants choosing to live in already-existing residential areas chose their home location based on a love for nature. Immigrants and natives (those born and raised in Teton Valley) chose the locations of their households for significantly different reasons. Immigrants chose their home sites primarily on the basis of natural amenities and economic considerations (e.g., jobs, or the cost-of-living). Few immigrants cited family or Teton Valley being where they were born as first-order considerations in their household-location decision. Conversely, natives cited family or Teton Valley being their home as their primary reason for the location of their households

Table 3.1 Primary reason that homeowners provided for choosing house locations in natural areas, agricultural areas, and residential areas in Teton Valley

| | Primary reason for house location (%) | | |
Group	Natural amenities	Economic constraints	Home place
Native residents	34	10	56
Natural-area immigrants	72	14	14
Agricultural-area immigrants	58	23	19
Residential-area immigrants	47	39	14

nearly twice as often as natural amenities, and they rarely cited economic considerations.

We found that environmentalism, educational level, and age predicted whether immigrants chose to live in natural areas, but gender and income did not. Immigrants moving to natural areas exhibited significantly higher NEP scores (57.46) than those moving to agricultural (53.43) and residential areas (51.59). Although the relationship between how environmentally oriented people were and where they chose to live appeared to be random at intermediate levels, the least environmentally oriented immigrants moved to natural areas at half the average rate and the most environmentally oriented immigrants moved to natural areas at double the average rate (fig. 3.3A).

Immigrants with graduate or professional degrees moved to natural areas at twice the rate of movement to agricultural areas, and 10 times the rate to residential areas. A less-pronounced version of this pattern occurred at the college graduate level, and a slightly lower one for immigrants with some college education. Immigrants with associates degrees or less preferentially chose not to live in natural areas (fig. 3.3B). Older immigrants tended to choose houses in natural areas (fig. 3.3C). This trend changed slightly for the oldest respondents (in their late 70s and 80s), probably because homes far from town in areas with zero public transportation were not ideal for people needing ready access to medical care.

We divided households into categories we called large (more than two people) and small (one or two people). Average household size in both the United States as a whole and in the states of Idaho and Wyoming hovered between two and three individuals per household for the last three decades, so the division seems intuitive. Respondents with small households were almost four times as likely to live in natural areas as respondents with larger households (fig. 3.4).

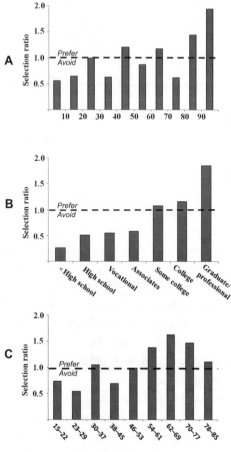

Figure 3.3 Newcomer preference or avoidance of homes in natural areas. The relationships are between the weighted percentage of those immigrants choosing to live in natural areas and (*A*) their new ecological paradigm (NEP) score, reported in deciles (9 values that divide the scores into 10 equal parts); (*B*) their educational level; and (*C*) their age.

In Teton Valley, house addiction reversed the typical positive relationships between environmentally sensitive behavior (e.g., recycling, watching nature-related television, or donating money to environmental organizations) and educational level, environmentally oriented attitudes, and pro-environmental behavior.[36] Household location, as an indicator of environmental behavior, yielded a negative relationship between age, educational level, NEP, and pro-environmental behavior. Older, highly educated immigrants with environmentally oriented attitudes frequently chose to live in natural areas (e.g.,

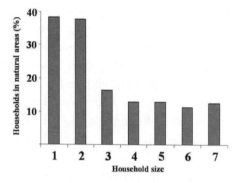

Figure 3.4 Household size (the average number of people per household) and the percentage of respondents choosing to live in natural areas.

riparian zones, wetlands, and critical winter range for wildlife), while immigrants with the lowest levels of education and least environmentally oriented attitudes often selected previously established residential areas. Natural areas were the most difficult locales to develop, due to permitting, but those with higher educational levels, and the most environmentally oriented immigrants, still choose to live in them. For most of these respondents (72% of those living in natural areas), household location was an explicit choice, motivated by natural amenities. Income did approach statistical significance in our results, and it remained significant if we did not control for educational level. This suggests that income may provide the means by which a person could choose to buy a home in a natural area, but factors associated with a college education supplied additional motivation for that choice.

These findings carry particular importance for other regions where household proliferation associated with natural amenities threatens the environment.[37] Although the potential cases regarding this phenomenon are too numerous to list fully, some examples are housing developments in rural areas with natural amenities that threaten bird diversity in the Puget Sound region in Washington, fish species in the Northern Highland Lakes District in Wisconsin, Key deer (*Odocoileus virginianus clavium*) in the Florida Keys, biodiversity conservation in the GYE, and panda (*Ailuropoda melanoleuca*) conservation in Wolong Nature Reserve in China.[38]

Our results suggest that a knowledge of natural or environmental science alone cannot address these conservation challenges. Many immigrants choosing to build homes in riparian areas or on hillsides in Teton Valley held advanced degrees in related fields, such as wildlife ecology, fisheries, zoology, and forestry. One biologist, a recent retiree from the Idaho Department of Fish and Game, pointed out this issue long before it was supported by our survey results. He spoke loudly, so his voice would carry over the nail guns that were tacking his new home together: "The biggest problem is the loss of winter range [for mule deer and elk], and I've now become part of it because my wife won't live in town." Educational efforts that highlight relationships between household-location decisions and their environmental impacts may increase understanding, but they will not necessarily trump house addiction.[39]

These findings demonstrate the need for an explicit consideration of household-location decisions based on resource use and conservation, and for the development of ways to experience pristine environments beside building houses on them. Linking household-location decisions to environmental attitudes represents a critical step for conservation issues. By themselves, environmentally conscious decisions about home appliances, food consumption, family planning, voting, donations, activism, and transportation alone will not protect the environment; people also need to make environmentally conscious decisions regarding the location of their homes.

Our Teton Valley study highlights several critical questions. First, how will an aging population influence household-location decisions and their ecological impacts? Preliminary research on this topic suggests that modern retirement communities may have unusually large, detrimental effects on the environment and on conserving biodiversity.[40] These effects, however, have yet to be quantified, or to be projected into the future. Second, how can we modify education to promote environmental attitudes and a coherent relationship between those attitudes and household-location decisions? Third, how can conservationists help create a positive relationship between environmental attitudes and environmentally sensitive decisions on household sites?

Fighting Community Planning in the Name of Property Rights

Our study suggested that the dynamics surrounding community planning in Teton Valley were as ironic as those surrounding where people chose to build their homes. The individuals building homes in natural areas were environmentally oriented and well-educated immigrants, and those adamantly opposed to change were vociferous opponents of the community planning that would have helped manage change. A possible explanation came from the tendency to naturalize property rights. For naturalized-property-rights believers, property represents a fundamental, inalienable, presocietal right that is superior to rules imposed by society. People may hate messy, unplanned development, but they don't act to prevent it, because they believe naturalized property rights trump their own preferences. Thus a belief in naturalized-property-rights values should be able to predict whether residents thought community planning should occur, and whether they were willing to engage in planning processes regardless of their attitudes toward development.

We tested this hypothesis by reviewing lengthy notes from our interviews with Teton Valley residents and separating them into individuals who believed in naturalized property rights and those who believed that property rights were political constructions. Naturalized-property-rights believers make statements such as "Everyone has a right to live where they want to" and "Everyone has a right to use their property as they see fit."[41] We also controlled for several variables other than naturalized-property-rights beliefs, ones that might explain why residents would want or would choose not to participate in community planning, including local status, religion, political affiliation, education, income, age, and gender.[42]

We identified four themes among the responses to the future of development in Teton Valley: remain rural (1%); grow rapidly (70%); become a wealthy, recreation-centered community, such as Aspen, Colorado (15%); and become a big, unplanned subdivision (13%). While all residents predicted future growth and loss of the valley's rural nature and 66% were unhappy with their own prognostications, only 18% thought that the future development they foresaw was a good thing. Respondents with naturalized-property-rights values and political-property-rights values were equally unhappy (naturalized-property-rights = 64%, political-property-rights = 65%) with their vision of future development. Although most residents were dissatisfied with their projections for future development, 61% said nothing should be done. Only 39% had suggestions for improvements. When we separated the responses into naturalized- and political-property-rights groups, the two camps were sharply divided in their opinions of what should be done to make future development better: 90% of the political-property-rights advocates provided a suggestion, while 87% of the naturalized-property-rights advocates stated that nothing should be done. A person's view of property rights was an almost perfect predictor of whether he or she thought community planning was needed.

Given their disparate views on property rights and community planning, most residents still shared the same concerns about development. The first was a sense of loss regarding farming, local culture, and access to natural resources. Respondents (designated by R plus a number) expressed concern over the disappearance of farming and the loss of farmland. "I don't like it [development]. I would like to wave a magic wand and make it all like it used to be" (R309). "Farming is done in the valley. I don't blame the farmers, though. We've had to sell some of our ground" (R259). Another asked: "You've been to Teton Springs [a farm recently transformed into a golf and fishing resort]? Oh, I hated to see that farmland go. . . . I'd have liked to save Game Creek [a tributary of the Teton River]. I hated to see the river go. Nothing can stop the development, but I wish we could, and bring back the creek" (R411). Still others were outspokenly hostile to development that pushed out agriculture. "They're going to turn it into what they did in Jackson[, Wyoming]. They are going to ruin it. They're trying to subdivide every square inch. They don't give a damn about the farming and ranching. The same goddamn people who developed Jackson are developing here. They come in saying 'We just want to fix it' but then they try to make it like where they came from" (R244).

Community members also focused on the loss of local culture. "It's [the valley's] going to pot. Subdivisions, golf courses, stuff we don't need. A different kind of people than who was raised here are taking over. I don't like it" (R233). They also assumed others shared their sense of loss. One explained, "Everyone prays for a nice quiet valley like the past. . . . This is what they call progress" (R63). Several expressed deep sadness that their valley had lost its familiarity. "I used to know everybody and their dogs, now I don't know a soul. . . . Glad my life is behind me" (R110). Finally, respondents expressed concern over their loss of access to natural resources, which was associated with the cultural shift. "There used to be a lot of places to go for good fishing, then people saw what we had and bought it up and put up No Trespassing signs. Now there's very few places for people to go" (R601). Their belief in naturalized property rights offered no protection from these losses.

The respondents' second concern was the hypocrisy associated with building a home and then preventing others from doing so. "Well, I moved here and added to the influx, so I can't stop others from doing the same thing, even though I would like to" (R9). "People move in here and crowd the place up. Everyone wants to be the last person who moved in" (R299). "I don't like it. It's going to become like the rest of America. . . . I moved here, so I'm part of the problem" (R149). "How can I throw stones? I came here. It's part of the problem. I don't want the Colorado syndrome where you slam the door after yourself" (R572). Both old-timers and new residents found themselves trapped by their own household behaviors. Preoccupied by fairness, they could find no rationale for denying anyone the pleasure of doing whatever they wanted with their private property.

Their third concern was frustration about "outside" control over land use. Most respondents linked this lack of individual and local control to their belief in the necessity of allowing market forces to dictate development. "Development is my main reason for moving. People won't let you on their land anymore. Money runs it, and there's nothing poorer people can do about it" (R321). Those in control were always an elusive corporate "they." For example, "It's gonna be a resort area. It already is. They are taking it away from us" (R266). And "they" are always wealthy: "In the end the rich seem to be the ones that . . . whoever has the money, they are going to do what they want. Everything is fueled by the almighty dollar" (R218).

At first glance, these concerns may seem like nothing more than idle grumbling. They do, however, provide a possibility for finding common ground between those who view property as a natural right and those who see it as a political right. Political-property-rights advocates mentioned the same topics in their discourses (concerns about local losses, the hypocrisy associated with home-building practices, and market control over development) as the naturalized-property-rights proponents did. One political-property-rights advocate said: "We all need to remember why we moved here. We like the small town, the roots, and heritage of the town" (R44). Several of these respondents expressed concern about the livelihoods of individual farmers, suggesting that "the community needs to compensate landowners in some way for not selling into subdivisions" (R518). "I don't like the way development is happening, but farmers need a way out. They worked hard and deserve to be paid" (R199). Some offered suggestions for specific agricultural sectors: "Big spud farmers are in a bind. We have to find alternatives for them. Financial incentives, tax benefits for conservation easements. . . . More than what's out there. Give them a higher density transfer [the ability to sell higher-density development rights to property owners in urban-proximate areas in exchange for not developing their agricultural land]" (R392).

These respondents also demonstrated concern over the hypocrisy associated with moving to an area and then prohibiting others from doing so. "Do you know the difference between a developer and an environmentalist? A developer wants to build a cabin, and an environmentalist has one. Like Ted, who started the land trust, or the folks in VARD" (R430). As one explained:

"I moved here, so I can't say anything if other people want to [move here]. I just wish they'd take care of it. And they don't need to put up No Trespassing signs. When I moved here, it was easy to get permission to hunt and fish. Now it's pretty hard" (R497). Finally, political-property-rights advocates identified market control over development as a threat: "We'll have houses from end to end. It's sad because most of the local people will have to move away. . . . We should try to change it with zoning, but it's inevitable, because the developers are so persistent" (R156). Another reflected that "unfortunately, this area attracts people with money and they will use it to get what they want" (R176).

Our results suggest that Teton Valley residents cared about landscape planning, but naturalized-property-rights values prevented many of them from acting on their concerns. These values were more predictive of respondents' beliefs regarding development planning than political affiliation, religion, local status, education, age, and gender.[43] The positive relationship between income and the belief that community planning should occur may relate to issues of real and perceived power. Indeed, since community members did not engage in planning, most planning was carried out by developers. By refusing to take part in community planning, naturalized-property-rights advocates gave exclusive authority to economic interests.

The power exerted by developers and investors over development in Teton Valley was made possible by the self-subjugation (the denial of their rights) of those valuing naturalized property rights. Respondents who considered that community solutions to development problems were impossible, and who gave exclusive importance to naturalized property rights, were opposed to future unplanned development. They told heart-wrenching stories about losing access to their favorite berry patches, fishing holes, and hunting spots; seeing their old horseback-riding trails cut up by subdivisions; and watching creeks dry up for the first time in memory. Naturalized-property-rights advocates held ultimate management power, but they wielded it against themselves. For them, being good citizens entailed disciplining their own desires and allowing others to do "whatever they want with their property." By naturalizing property rights, these residents demonstrated how even democratic self-governance can be dangerous.[44] Their tendency to "frantically hide behind unhistorical and abstract universalisms" (e.g., naturalized-property-rights) ultimately denied them the political and moral choices they needed to make their own decisions about the future sustainability of their community.[45]

Our results uncovered three themes capable of fostering the dialogue needed for community planning in this context: (1) the value of past culture and previous land use, (2) the ethical problems associated with building a home and then denying others the same right, and (3) the excessive influence of market forces on development. These themes emerged from the statements of both naturalize-property-rights advocates and political-property-rights advocates, and the themes functioned in concert with each other. For example, when expressing sadness about the loss of local culture, a respondent may implicate him- or herself as one of the newcomers who has contributed to this loss, and sigh over the power of wealthy developers from outside the valley.

These themes provide points of tension within the naturalized-property-rights story, and an opportunity to encourage naturalized-property-rights advocates to think about the implications of their views. They may help clear a space for productive debate. And that debate may create a rupture in the ideology that prevents naturalized-property-rights advocates from engaging in community planning as a means of acting on their concerns about development. For residents who believe that "we need to work with the community we want to keep instead of alienating the community we need to work with" (R268), the themes represent common ground for collaborative land-use planning.

Opportunities for Rebuilding after the Housing Crisis

Perhaps the first lesson learned from Teton Valley is that communities must apply the same foresight to community planning that they apply to economic development. Had the residents considered where all the houses would go during the same 1964 town meeting when they decided where the local economy should be heading, they would have been better prepared when speculators descended on the small community with duffel bags of cash and plans to build homes helter-skelter across the landscape. Without a preexisting framework outlining where housing development should occur, and where it should not occur, planning and zoning decisions that limited development appeared capricious and in violation of property rights. Further, those hoping to preserve the area's rural character and valuable ecosystem services (e.g., blue-ribbon trout streams) did not have the time or the resources to make their case in the face of the overwhelming demand for development. Designing regulations that favor development in areas that promote a sustainable tax base, walkable communities, rural character, and clean water—and that restrict development in areas that hurt community interests by generating unsustainable tax burdens, eliminating agriculture, and fouling trout streams—would force future development booms into a structure that eventually leads to what residents want without appearing to single out individual landowners.

Unfortunately, we also learned that with few exceptions, house addiction will trump environmentalism and altruism. In the absence of enforced regulations or market mechanisms making development impossible in pristine natural areas, even the greenest people alive will build houses in them. It wouldn't hurt if the most vocal critics of sprawl got some roommates and moved to an apartment above a store downtown, but that is not enough. Rather, communities must rely on regulations that either ban development in critical areas or extract enough tax monies from that development so damages can be offset elsewhere through expensive mitigation.

These changes, however, must be implemented in a way that shows deference to property rights, and that uses local values to place boundaries on what is considered a property right. In Teton Valley, rural character, outdoor recreation, and local control were important cultural values that could be tapped to constrain absolute visions of property and to provide buffers for otherwise unrestricted market control. Restrictions on unplanned growth could be argued for on the grounds that such growth undermines current residents' rights to recreation, clean water, and self-determination. A debate between advocates of local control and outside developers hoping to use property rights to squelch local input on development decisions would have been more productive and more realistic than the somewhat contrived divide between farmers hoping to cash in on their landholdings and hypocritical environmentalists.

The Great Recession offered unprecedented opportunities both to learn from the past and to rectify past planning mistakes. It created an opening for smart growth in Teton Valley by forcing house addicts to get clean by going cold turkey. It wiped out their bank accounts and took away their second homes and investment properties with a flood of foreclosures. The destruction of household wealth, credit contractions, and employment instability did away with many consumers' ability to purchase homes, particularly homes in resort communities.[46] This facet of the Great Recession is likely to persist for more than a decade. Further, real estate professionals in resort communities suggest that the real estate collapse has created a generation of home buyers with some immunity to the rationalization that homes in these areas are a safe and lucrative investment.

Teton Valley residents, and VARD members in particular, have seized this opportunity and started taking back control over development with research and capacity building (the means to achieve a sustainable result). They helped form a joint venture with the Sonoran Institute and the Lincoln Institute of Land Policy to reshape development patterns in Teton Valley. The project's kick-off meeting in November 2009 involved assessing the challenges and opportunities left in the aftermath of the real estate crash, and considering solutions to the premature and obsolete subdivisions filling the valley. The effort started with legal, economic, and ecological research.[47] This research laid a foundation to keep the county out of future lawsuits associated with regulating development. It documents when local governments have authority, what constitutes procedural due process, what gives owners a vested right to develop their property, and what constitutes a taking of property (the government exercising its right of eminent domain). This research puts a dollar figure on the tax burdens created by subdivisions located far from firefighting, school, sanitation, and police services. It also spells out what the ecological damages are that can be caused by developments in riparian areas and on hillsides. Even if development pressure doesn't build for a decade, these guidelines and their associated data will lift the burden of proof from local communities and place it on developers trying to sell plans that violate community goals.

VARD has also worked to harness market forces to promote smart growth by adopting the nascent field of ecosystem-service stacking. Ecosystem services (the benefits people obtain from ecosystems) can be stacked (with one ecosystem providing several ecosystem services), and payments or credits for these stacked ecosystem services allow landowners to receive multiple revenue streams for engaging in socioecologically sensitive planning. For instance, if a developer reduces the total number of planned lots and consolidates the remaining lots on less-sensitive land, while permanently protecting sensitive and ecologically valuable riparian areas on the rest of the parcel, that developer could receive payments for preserving riparian habitat, carbon credits, water quality credits, and conservation banking credits. VARD is piloting an approach called facilitated plat redesign that uses ecosystem-service stacking as an incentive to responsibly redesign failed subdivisions.

The first pilot project involved a failed development named Targhee Hills Estates. The plans for Targhee Hills Estates called for an exclusive gated retirement community surrounded by artificial water features. The proposed development bordered a natural water feature, Teton Creek, and had valuable water rights to create the proposed fountains and pools. Before the real estate bubble burst, developers invested $2 million in sewers and sold 18 lots. Complete lack of demand for the remaining lots left the developer on the verge of bankruptcy, a condition also threatening projects elsewhere in the United States. The existence of other developments in Teton Valley that had not been utter failures meant that the developer had some incentive to salvage the project instead of just walking away. It also meant that the county had an incentive to help the project along, thereby avoiding future infrastructure burdens. Any move forward, however, required the 19 lot owners (including a contractor) and two banks with promissory liens on the property to agree on redesign plans. VARD began the process by convening experts from the Teton Regional Land Trust, Trout Unlimited, and Friends of the Teton River to explore potential easement values and tax breaks associated with putting water back into natural streams, preserving open space, and protecting threatened species as part of the redesign process. They also explored the economics associated with leasing old barley fields back to farmers, and evaluated potential changes that would reduce building costs.

Stacking ecosystem-service credits as an incentive for sustainable developments is a step in the right direction for Teton Valley, but most likely it won't suffice when the house addicts return and houses start selling again.[48] The successes in 2009 relied on a complete lack of demand for houses, large investments of cheap or voluntary labor among planning advocates, and desperation among developers. When real estate transactions bounce back, local governments will need a stick to go with the carrot. Developers protecting open space should be rewarded, but those who threaten community interests (including open space, water quality, and wildlife conservation) should face higher costs and more onerous permitting processes. As long as local governments make the case that they are attempting to promote a legitimate governmental purpose, and give landowners some reasonable economic use of their property, their efforts should withstand legal challenges from developers wielding property-rights arguments.[49]

The final lesson learned from Teton Valley is that smart growth is incredibly complex, even without vocal opposition from developers or from residents willing to deny their own preferences in the name of property rights. During a 2010 visit to Teton Valley, we discovered that threatened Yellowstone cutthroat trout had made a miraculous recovery since the global recession hit. State wildlife managers had imposed strict regulations on fishing for cutthroats while declaring an unlimited open season on their competitors, rainbow trout and brook trout. Those regulations had existed for several years, without initial apparent effect. The recession correlated with an unusually cold and wet winter, at least in recent years, yielding more water in the streams. The recession also coincided with the end of a huge real estate boom that transferred water rights from farmers, who previously used it for sprinkler irrigation with high evaporation rates, to subdivisions with no houses. Water owned by empty subdivisions simply ran back into the rivers, and spring floods washed layers of sediment from gravel spawning beds, both of which aided the cutthroats' recovery.

No one anticipated these events, and no one knows for sure how much development patterns factored into the current outcome. If explosive speculation and development had never occurred, irrigation would have continued, and the creeks and rivers would still be starved for water. What is obvious is that current contexts provide opportunities to implement regulations—such as requiring native or xeric landscaping, and restricting irrigation—before the subdivisions fill with residents and the residents grow accustomed to turf-grass lawns and wasting water through a variety of consumptive practices. Since the people moving in are attracted by amenities like Yellowstone cutthroat trout, it should not be a hard sell.

Conclusion

Unfortunately, even in the best-case scenario, the scars of the housing crisis will remain in Teton Valley and the rest of the GYE. The sweeping vistas of ranchettes and subdivisions will not be erased by smart growth, community planning, or the extensive use of conservation subdivisions. Nonetheless, the GYE is a flagship for parks and natural areas facing unsustainable household dynamics globally.[50] The directions taken by activities around the Yellowstone area are dynamic, but even when conservation-development measures are attempted, trends toward growth in riparian areas highlight the ecological dangers posed by various forms of development, especially house addiction, and the need to defuse the housing bomb.[51]

CHAPTER FOUR

Household Dynamics and Giant Panda Conservation

Human impacts on the environment are common, even in many of the world's approximately 134,000 protected areas (accounting for roughly 13% of Earth's land surface).[1] Although protected areas are the cornerstone of biological conservation and are often perceived as the safest preserves for nature, household development still occurs in many of them.[2] In this chapter, we focus on household impacts in a flagship protected area, the Wolong Nature Reserve in China, established for the world-famous, endangered giant pandas (*Ailuropoda melanoleuca*). We begin with some background information about the study area and a conceptual framework. We then illustrate how household dynamics influence efforts to save giant pandas. Finally, we present promising approaches to reduce household impacts on panda habitat.

Characteristics of the Study Area

Wolong Nature Reserve (fig. 4.1) is in Sichuan Province, in southwestern China (30°45′–31°25′ North, 102°52′–103°24′ East). It was established in 1963 as a nature reserve and expanded in 1975 to its current size (200,000 hectares [494,210 acres]) to conserve giant pandas. It is one of the largest homes for these pandas and contains about 10% of the wild panda population (approximately 140, according to the most recent panda census in 2000).[3] Located between the Sichuan Basin and the Qinghai–Tibet Plateau, Wolong has a complex topography. Elevations range from 1,200 to 6,525 meters (3,937–21,407 feet), with high mountains and deep valleys, several climatic zones, and rich diversity of habitats.[4] Of the more than 6,000 animal and plant species within the reserve, 13 animal species and 47 plant species are on China's national protection list.[5] Further, Wolong is part of the international Man and Biosphere Reserve Network, as well as a world heritage site and a global

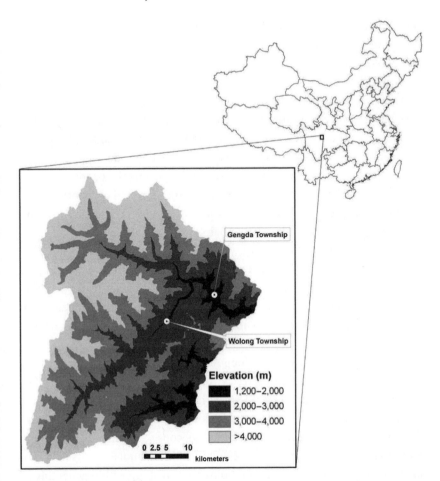

Figure 4.1 The location of and elevations in Wolong Nature Reserve. Chen, X., Lupi, F., He, G., Ouyang, Z. & Liu, J. Factors affecting land reconversion plans following a payment for ecosystem service program. *Biological Conservation* **142**, 1740–1747 (2009).

biodiversity hotspot.[6] As a flagship nature reserve, Wolong has obtained exceptional financial and technical support from the Chinese government and international organizations.

Like many other protected areas, there are human residents within Wolong. Although having households located inside nature reserves is rare in developed countries, it is common in China and many other developing nations.[7] In such countries, the vast majority of nature reserves were established in the past three decades, but many people had already settled in those areas for generations and centuries.[8] Local residents inside Wolong are mostly farmers and maintain a subsistence lifestyle. Diverse human activities occur within the nature reserve, including farming, fuelwood collection, timber harvesting, house building, the collection of Chinese herbal medicines, and road construction.

Although electricity is available in Wolong, it is relatively expensive, has unstable voltage levels, and frequently fails.[9] Thus local residents continue to gather fuelwood for cooking their own meals, cooking fodder for pigs (a major form of livestock in the reserve), and heating. Fuelwood collection sometimes takes place in the same forests inhabited by the pandas, where it destroys panda habitat, as these animals require a forested overstory, a bamboo understory, gentle slopes, and a suitable altitude.[10] There are two staple food sources for the pandas: arrow bamboo (*Bashania fangiana*) and umbrella bamboo (*Fargesia robusta*), which are both understory species. Thus tree removal—primarily for fuelwood, but also for house construction—is among the most important factors affecting panda habitat in the reserve.[11] Home building and residential land use may also affect panda habitat. Panda habitat can recover as the forest regenerates, but this recovery may take 40 years or more.[12]

Wolong is managed by the Wolong Administration Bureau, which reports directly to both China's State Forestry Administration and the Forestry Department of Sichuan Province.[13] Under the Bureau, there are two township governments: Gengda Township and Wolong Township (fig. 4.1). Each township includes three villages, and each village consists of three to six groups. A group is the lowest administrative unit in Chinese rural areas, usually consisting of a dozen to several dozen households that are within geographical proximity. Wolong has some unique characteristics (e.g., giant pandas), but it also shares common features with many rural areas in China and other countries, including a populace with subsistence lifestyles.[14]

Conceptual Framework

The interactions between households and panda habitat are complex (fig. 4.2). They can be shaped by human factors, such as governmental policies, and natural factors, such as earthquakes and landslides.[15] The human factors include population size / households, attitudes, needs, and activities. Forests have many critical attributes, including structure, function, quantity, and quality. When forests change, panda habitat changes, as the former is part of the latter and provides the food and cover that pandas need for reproduc-

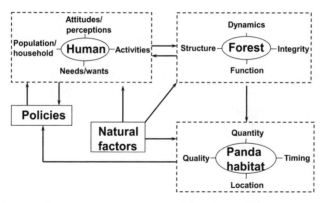

Figure 4.2 A conceptual framework of the interactions between humans and panda habitat. Liu, J. et al. A framework for evaluating effects of human factors on wildlife habitats: The case on the giant pandas. *Conservation Biology* 13, 1360–1370 (1999).

tion and other activities. The suitability of panda habitat is determined by abiotic and biotic (including human) factors.[16] Slope and elevation are major abiotic factors, as pandas prefer flat areas or gentle slopes where they can move around easily.[17] The main biotic factors include their primary food, bamboo, and forests (primarily deciduous broadleaf forests, conifer forests, and mixed conifer / deciduous broadleaf forests), which create the necessary growing conditions for bamboo and living conditions for pandas. Approximately half of the reserve is unsuitable habitat for the pandas, even without human impacts, because some regions are above the tree line and other areas have slopes that are too steep or elevations that are too high.[18] Changes in forest cover influence the quantity, quality, and spatial and temporal location of panda habitat in the reserve.[19] Through human activities (such as gathering fuelwood), households also directly affect forests and panda habitat.

Governmental policies influence households directly, and forest and panda habitat indirectly, through governmental regulations and financial support from central and provincial governmental agencies (e.g., China's State Forestry Administration and Sichuan Province's Department of Forestry). Both regulations and funding impact the Wolong Nature Reserve Administration Bureau, the townships, the villages, and (ultimately) the households and the local residents. These policies, however, can also be affected by local residents and by the conditions in panda habitat. Local residents sometimes do not comply with governmental policies when financial support from the government is not sufficient to meet their needs and wants. And habitat conditions have prompted the government to develop and implement more effective policies for conservation.[20] Moreover, human activities may be constrained by feedback from the forests. As forests containing fuelwood shrink and become more distant from households, fuelwood collection becomes more difficult.[21] Households may eventually have to adopt a different lifestyle, one without the use of fuelwood.

Household Proliferation

The size of the human population in Wolong increased from 2,560 residents in 1975 to 4,550 in 2005, while the number of households jumped from 421 to 1,156 during the same time period. Thus the number of households grew more than twice as fast (a 174.6% increase) as the number of people (a 77.7% increase). The more rapid expansion of household numbers compared with population size in Wolong led to an analysis of population growth and household proliferation in 141 countries, which revealed a similar pattern to the one in Wolong.[22] Globally, the number of households has been growing faster than the population, and the differential is highest in biodiversity hotspots.

Reasons for Household Proliferation

Household proliferation is a product of an expansion of the population and a reduction in the number of people sharing a home. The rapid population increase in Wolong was facilitated in part by an exception to China's "one child policy" that allows minority ethnic groups, such as Tibetans (who make up the majority of residents in the reserve), to have multiple children.[23] The primary reason for the upsurge in household numbers in Wolong, however, was the decrease in the number of people per household (fig. 4.3), as it is in many other places around the world.[24] The reduction in household size was driven by aging, the formation of new households by young people, and governmental policies. In this case there was only a relatively small number of immigrants, as the only legal way to immigrate into Wolong was through marriage.[25]

The human population in Wolong is aging rapidly. From 1982 to 1996, the number of people 60 years old or older increased by about 25%, while the total population rose by less than 15%.[26] Because older residents lived in homes with fewer occupants, this expanding demographic contributed to the

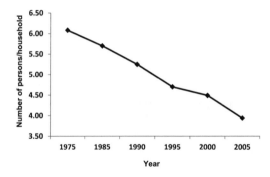

Figure 4.3 The dynamics of household size in Wolong. Liu, J. et al. Beyond population size: Examining intricate interactions among population structure, land use and environment in Wolong Nature Reserve (China), in *New Research on Population and Environment* (ed. B. Entwisle and P. Stern), 217–237. National Academy of Sciences Press, 2005.

proliferation in households. This proliferation also resulted from shifting behavior among Wolong's younger residents. Like many other locales in China, young people in Wolong are now more likely to form their own households, rather than live with their parents and grandparents under one roof, as was traditionally done in China.[27] There are many factors affecting young people's decisions to leave parental homes, summed up in their concerns about "less conflict with siblings," "more independence," and "we have to leave because of too many siblings." Parents also had significant influences on the adolescents' decision-making process.[28] Parents in Wolong began to see their children leaving the parental home as a new norm: "It is a normal phenomenon, as a tree would have many branches when it grows up."[29]

Governmental policies had profound (although unexpected) impacts on household formation in Wolong. There was a sudden and unanticipated addition of 65 new households in 2001, almost three times higher than the average annual rate of increase during the previous 25 years (approximately 21 new households a year from 1975 to 1999).[30] These extra households were formed to take advantage of subsidies from the Natural Forest Conservation Program, which was started in 2001 to protect natural forests from illegal harvesting.[31] Households were assigned to monitor and protect forest parcels, and the government provided subsidies for this on a per household basis. The subsidies were large, and they accounted for approximately 20%–25% of the total annual income for most of the households involved.[32]

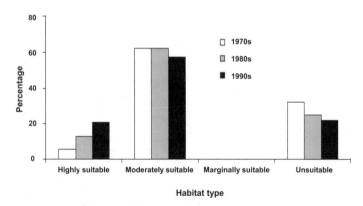

Figure 4.4 Fuelwood collection sites in three decades, divided into four types of panda habitat. He, G. et al. Spatial and temporal patterns of fuelwood collection in Wolong Nature Reserve: Implications for panda conservation. *Landscape and Urban Planning* **92**, 1–9 (2009).

Impacts of Household Proliferation

Household proliferation in Wolong has a range of socioeconomic and environmental impacts. Greater numbers of households require more land for house sites, more timber for house construction and furniture, and more fuelwood for heating and cooking.[33] As the total amount of fuelwood consumption increases, residents exhaust that resource in the forests near their homes and are forced to spend more time gathering fuelwood in more distant areas. As a result, the average distance between homes and collection locations increases over time.[34] The forested areas farther from households provide the best panda habitat, so as fuelwood collection expands into those areas, it creates progressively more damage to suitable panda habitat (fig. 4.4).

A reduction in the number of people sharing a home not only increases the number of houses, it promotes greater fuelwood consumption per capita (fig. 4.5). Households with three people use about twice as much fuelwood *per capita* (5 cubic meters [177 cubic feet]) as households that contain six people (2.5 cubic meters [88 cubic feet]). This is because the number of people living in a home has relatively little impact on the amount of fuelwood needed to heat the building. Houses that have fewer people in them require just as much fuelwood for heating as houses that contain more people. Aging also influences fuelwood-collection patterns, because the heating season for the elderly is longer than that for younger people (starting earlier and ending later), and households with elderly residents use more fuelwood for heating

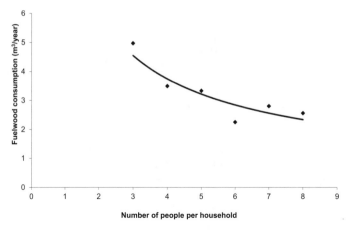

Figure 4.5 The per capita fuelwood consumption in different household sizes. Liu, J. et al. Beyond population size.

than those without seniors.[35] For cooking, more fuelwood is needed in a large household, because more food needs to be cooked for a greater number of people, but the efficiency per capita is still higher in a larger household.[36]

Household proliferation has contributed to significant losses of suitable panda habitat in Wolong. From 1965 to 1997, both forest cover and panda habitat in Wolong were dramatically reduced (fig. 4.6), because people liked to collect resources (e.g., fuelwood and timber) in areas the pandas need. Both people and pandas prefer locales that are not too steep and are below the tree line. The most suitable panda habitat has become more fragmented as human activities (e.g., gathering fuelwood, harvesting timber, constructing roads, and building homes) split it up (fig. 4.7). Simulation using a spatially explicit model (one focusing on spatial distribution) indicates that the existing levels of household formation and their associated fuelwood consumption led to an additional 10% loss in panda habitat at elevations below 2,600 meters (8,530 feet), compared with hypothetical conditions in which there were no additional households after 1997.[37] In the model, decreasing household fuelwood consumption by two-thirds reduced the loss of habitat below 2,600 meters by 59%, compared with baseline scenarios.

The quantity of panda habitat is more sensitive to factors influencing household numbers than to those affecting human population size.[38] Simulations using an agent-based (multiagent) model produce very different results in the numbers of households versus the numbers of people when variables change, including age at one's first marriage, the interval before the first child

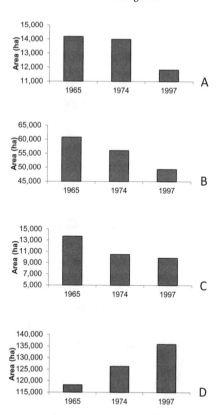

Figure 4.6 Amount of panda habitat in Wolong Nature Reserve before and after the reserve was established in March 1975: (*A*) highly suitable habitat, (*B*) suitable habitat, (*C*) marginally suitable habitat, and (*D*) unsuitable habitat. Liu, J. et al. Ecological degradation in protected areas: The case of Wolong Nature Reserve for giant pandas. *Science* **292**, 98–101 (2001).

is born, the amount of time between births, the number of children, and upper birth age (the oldest age for women to give birth). Fertility-related factors (e.g., the fertility rate, the birth interval, and the upper childbearing age) lead to almost instantaneous changes in human population size, but they have time lags of 20 years or more before they impact the number of households.

Adjustments in a person's age at first marriage produce the quickest change in the numbers of households. A marriage-age reduction from 38 to 18 years old could create a 10% jump in household numbers within five years. This difference is largely due to the household life cycle: delayed marriage postpones both the formation of new households and the births of babies who may form their own households in 20–30 years. It takes longer for other factors to

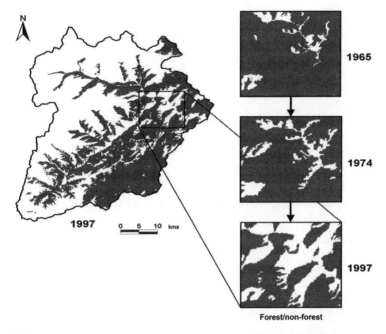

Figure 4.7 The forest distribution pattern across Wolong Nature Reserve in 1997 (*left*), with illustrations of the loss and fragmentation of the forest within a representative area before (*top right* and *center*) and after (*bottom right*) the current boundaries of the reserve were established. Gray areas are forested; white areas are nonforested. Liu, J. et al. Ecological degradation.

influence the number of households. For instance, when the fertility rate increases by one child per family, the children still stay with their parents until they establish their own households, creating a 20-year time lag before any change occurs in the number of households. These findings have important policy implications. If the goal is to reduce population size, lowering the fertility rate or the upper childbearing age would be most effective; to reduce the number of households, promoting marriage at older ages would create the best results.

Lowering the number of households by raising the age at which a person is first married is the fastest and most effective way to reduce the loss of panda habitat. Panda-habitat dynamics are affected more by the quantity of houses within that habitat than by the size of the area's human population. In part this is due to the way that fuelwood is consumed. A large proportion of fuelwood is used for heating, which changes very little if one person is added to or removed from the existing household. For cooking, the extra fuelwood used by adding one person or the fuelwood saved by removing one person would be small.[39] It takes over 20 years for a reduction in the fertility rate to begin protecting panda habitat, while changes in the age at one's first marriage have shorter-term impacts.

Effects of Governmental Policies

Policies not only affect the number of houses in Wolong, they also have an impact on how households there use resources (e.g., fuelwood) that critically alter the areas where the pandas live, as well as on where those households are located on the landscape. Policies that promote electricity use in the nature reserve lower the householders' dependence on fuelwood gathered from its forests and thus help protect giant panda habitat in Wolong.[40] The probability of residents switching to electricity is heavily affected by price, but the quality of the electrical supply (the outage frequency and the voltage variability) also has significant effects on the probability that householders will change over from fuelwood to electricity. A reduction of 0.05 Yuan per kilowatt-hour (from 0.13 Yuan per kilowatt-hour) would double the number of households using electrical power for cooking and heating. Yet a switch to electricity is also dependent on having few power outages. Thus both a low price for electricity and high-quality delivery are required to reduce fuelwood use and protect panda habitat. To reduce the price of electricity and increase its quality, Wolong Nature Reserve has built an eco-hydropower plant that now offers a more reliable electrical supply to many local residents. The Natural Forest Conservation Program and the Grain-to-Green Program have provided subsidies to farmers so they can now afford to buy electricity, thus leading to panda-habitat recovery.[41]

Historically, households in Wolong were widely dispersed across the landscape. New households are now increasingly located in areas close to the main road and near tourism centers, to take advantage of the tourism business and convenient transportation. In the 1980s, the Chinese government and the World Food Program built a large apartment complex at a location convenient for humans and distant from panda habitat. They hoped to lure farmers away from core panda habitat and into this apartment complex. The move and "upgraded" housing were offered for free, but no farmers made the switch.[42] In hindsight, policymakers realized that farmers could not move to housing that was geographically isolated from their croplands.

The deadly 2008 Wenchuan earthquake destroyed infrastructure over a large area, including Wolong, but it provided a unique opportunity to move households off of steep slopes, and away from panda habitat, to areas near the road. Wolong was close to the epicenter of the Wenchuan earthquake that, as of September 2008, left 69,227 dead, 17,923 missing, and 374,643 injured.[43] Inside the Wolong Nature Reserve, 148 deaths were reported as a result of the earthquake.[44] The earthquake also extensively damaged the main road and most buildings. In response, the government developed a variety of programs to stimulate economic recovery.[45] Massive infrastructure-development projects are being implemented (e.g., more than 300 million U.S. dollars are being invested by the Chinese government and Hong Kong).[46] Plans were created to stimulate the recovery of the reserve itself after the earthquake. These plans include the relocation of households, especially from areas that suffered heavy infrastructure damage due to earthquake-associated landslides, to areas with a lower susceptibility to landslides (e.g., with gentle slopes); the reconstruction of damaged facilities; and the promotion of tourism, which became virtually nil after the earthquake.[47] Based on these plans, as many as 75% of all households may be relocated to relatively flat areas within Wolong.[48] Participation in this relocation is voluntary, and is dependent on the level of damage to houses. When the plans are completed in 2015, the relocations will increase the average distances between households and panda habitat.

Providing incentives for the children of local farmers to emigrate may prove to be the most effective long-term solution for protecting panda habitat in Wolong from encroaching housing. Investing in the education of local farmers' children will ultimately help them attend universities or technical schools, find jobs, and settle down in cities, where most of the economic opportunities in China are centered.[49] Previous efforts to subsidize the relocation of local farmers were largely unsuccessful, because many farmers did not want to move. Even among those who relocated, many returned to Wolong after staying outside the reserve for some time. Although the local farmers themselves are not willing to move, due to such reasons as the lack of land and an inability to adapt to a new environment (e.g., temperatures outside the reserve are much higher than in Wolong), most farmers want their children to have opportunities for higher education and to find jobs in cities.[50] More importantly, research indicates that the young people themselves want to capitalize on educational and job opportunities outside Wolong.[51] This is significant, because moving one young person out of the reserve is equivalent to relocating several older people, since the young are the ones to establish new households and have children who build their own homes.

Therefore, it is both more socially acceptable and more ecologically effective to provide better educational opportunities for young people than to relocate older farmers.[52] This goal was facilitated by the need to temporarily move school-age children out of Wolong to continue their schooling after the earthquake, since this exposure to an outside environment may further increase the chances for these children to find jobs elsewhere.

Conclusion

Wolong Nature Reserve offers a laboratory for studying household proliferation and its relationship to wildlife habitat. Research in Wolong suggests that household proliferation and its drivers are more profoundly affecting panda habitat than human population size per se. Despite the enormous negative impacts of household proliferation, they can be minimized through appropriate policies and incentives. Focusing on improving educational and employment opportunities for young people would benefit both Wolong's human residents and its giant pandas. This finding highlights the fact that housing is part of a complex system. Although promoting multigenerational households may alleviate resource demands on a large scale, encouraging the youth of Wolong to leave home and seek educational and employment opportunities elsewhere is an environmentally responsible strategy at the scale where panda conservation is occurring. Wolong serves as a reminder that housing-related policies benefiting both local people and wildlife conservation have better chances of long-term success than less socially just practices, such as creating conservation refugees by evicting people from protected areas.[53] Although the Wenchuan earthquake was devastating to Wolong, the ensuing conservation and relocation programs have provided a unique opportunity to remodel the communities within the reserve in ways that reduce human vulnerability to natural disasters and increase the chances of survival for giant pandas. The case of Wolong suggests that household dynamics drive resource use, land use, and biodiversity conservation in diverse areas beyond the United States, and that the creative remodeling of homes, home locations, and household dynamics holds keys to improving humans' quality of life and promoting sustainable ways of life.

CHAPTER FIVE

DEFUSING THE HOUSING BOMB
WITH YOUR HOUSE

Households are the nexus for energy use, natural-resource consumption, and waste production around the world, despite being incredibly diverse in size, shape, and function. This chapter focuses on how to make U.S. households more sustainable. Challenges to sustainable housing that can be experienced worldwide often occur in their most extreme form in the United States: the large physical size of houses, few residents per unit, infrequent use of multi-unit housing, poor insulation, reliance on fossil fuels for heating and cooling, transportation and services provided by personal motor vehicles, and sprawl. Household energy consumption, including non-business travel, accounts for 38% of the total amount of energy used in the United States, and that portion has steadily increased from its 30% share in 2000.[1] Energy use by U.S. households alone is greater than the total energy use of any nation in the world except China. Energy consumption directly related to households exceeds energy use in the industrial sector (32.5%), the commercial/service sector (17.8%), and non-household transportation (11.7%). Because most of the energy consumed in the United States is derived from fossil fuels, household energy use also constitutes the largest and most rapidly growing sector contributing to climate change. In developed nations, the energy used by households is supplied primarily through fossil fuels, although there is a growing reliance on nuclear power in some nations, such as France; energy demands in developing nations more often directly rely on burning wood.[2] The collection of fuelwood can threaten endangered species, such as giant pandas,[3] and cause rampant deforestation, as has occurred in Haiti and some African nations. Fortunately, the vast majority of householders in the United States want to reduce their energy consumption, and they can cut their usage by 30%–50% without sacrificing money or comfort.[4]

Natural-resource consumption in households, however, is not limited to the resources needed to provide energy. Homes are constructed from various other natural resources (e.g., copper, iron for making steel, and wood), and the infrastructures needed to support households (e.g., roads, power grids, and sewer systems) require additional resources. Furthermore, households are where we consume food and dispose of our wastes. There were 389.5 million tons of municipal solid wastes generated in the United States during 2008.[5] Although the amount of household garbage has declined from a high of 413 million tons in 2006 (prior to the Great Recession), similar improvements have not occurred for transforming waste into energy through municipal garbage-burning programs. In 2008, the percentage of municipal solid waste converted to electricity was the lowest on record (a full percentage point below the 8% recorded in 1989).[6] The three largest contributors to the waste stream (paper, food scraps, and yard trimmings) are all organic and 100% recyclable. Recycling rates for paper (72%) are improving, but the reuse of yard trimmings (57%) and food scraps are not. In 2010, food scraps made up 34% of all discarded waste, and yard wastes constituted 14%. Well over 50% of the discarded waste is organic and, as it decomposes in landfills, produces methane, a greenhouse gas 20 times more powerful than CO_2.[7] Methane from landfills accounted for 22% of all U.S. anthropogenic (related to or resulting from the influence of humans) methane emissions in 2008.[8] This material could be converted to mulch by householders or municipalities, burned for electricity, or converted to biogas, all of which could make a profit and reduce the methane production of landfills. The steady stream of yard wastes into landfills prompted 25 states to ban these materials by 2008.[9]

Householder decisions also drive urban ecology on a massive scale. By 2050, over 8% of the United States will be covered by urban areas.[10] The collective decisions of householders form unique watersheds with dangerously erratic water flows that bounce between huge floods and dry streambeds; create erosion problems and high levels of pollutants; shape massive landscapes with little or no value to native wildlife; alter microclimates, typically by making urban areas hotter; and change patterns in the cycling of carbon, sulfur, nitrogen, and phosphorus.[11] Despite native-plant-based landscaping being cheaper to install and maintain, providing better wildlife habitat, helping to reduce erosion, stabilizing nutrient cycles, and often being preferred over turfgrass by urban residents, it is rarely put in place for new homes or as part of landscaping retrofits.[12] More native species are endangered by housing, or by invasive plant and animal species introduced by householders, than any other cause.[13]

If these switches are so easy, save so much money, and follow people's desires to make household decisions that help the environment, then why aren't they being made? Some changes, such as residential biogas production, are being deployed rapidly. In places like China and India, millions of households are already equipped to produce their own biogas from household and livestock wastes. Most improvements, however, are painstakingly slow. One reason is that the less sustainable households are, the more financial gain there is for those who profit from providing energy and potable water, and managing waste. Regulatory changes decoupling profits from the quantity of energy sold or the amount of waste removed are a logical first step in solving this problem, at least at the macro (large) scale.[14] In this chapter, however, we focus on what individual householders can do without relying on governments, and on how individual householders can overcome barriers to action. We discuss the most important ways residents can reduce their energy use, their waste production, and the ecological damage associated with their homes, and we pay particular attention to how structural and behavioral impediments to the suggested actions can be removed or overcome.

Sustainable Household Energy Use
Saving Energy

The key to energy sustainability in the residential sector is reducing use, not increasing production. Energy-efficiency measures can cut 23% from projected energy use in the United States, and reduce energy consumption by 9.1 quadrillion (15 zeros) BTUs from current projections for the year 2020.[15] Households offer a greater potential for energy-efficiency savings than the commercial or industrial sectors, because household energy consumption per unit of floor space has decreased by only 11% between 1980 and 2008, while efficiency has improved twice as fast in the commercial sector (21%), and four times as quickly in the industrial sector (41%).[16] This discrepancy reflects the fact that industry and businesses can save enough money through energy-efficiency measures to pay for an employee who determines the best actions to take, contracts for the work, and ensures its eventual completion. Average householders procrastinate, and they face diverse activities competing for their attention, so they rarely make energy-saving changes, even ones that produce financial benefits the moment they are implemented.[17]

Households have fallen so far behind in energy efficiency that, without considering household transportation, they hold 40% of the potential for improving energy efficiency and a majority of the potential for energy savings when transportation is included.[18] Just over half of the energy use associated with houses is consumed in the home (57%), while the remainder (43%) is used traveling to and from homes. Householders who can afford energy-efficient improvements have a huge array of options. Several groups, including the U.S. Department of Energy, have attempted to address the challenge of too many choices for energy efficiency by publishing relatively short, ranked guidelines for how householders can strategically reduce their energy use, but even these guidelines are relatively complex.

There are five categories offering possible energy savings in homes: replacing lighting and major appliances (11%), replacing electrical devices and small appliances (19%), building new homes (10%), retrofitting existing non-low-income homes (41%), and retrofitting existing low-income homes (19%).[19] Householders in low-income homes have limited means to achieve energy efficiency. One remedy would be to either subsidize improvements directly or guarantee loans for the improvements that are needed. Such programs could come with virtual loan guarantees to minimize risk. Loans could be made for improvements, but utility bills would be locked in at preimprovement levels. Householders would continue paying the same amounts they previously had for utilities, and the savings—that is, the difference between the sums paid and the lower cost of their actual utility use because of the energy-efficiency improvements—would be applied toward the loan, until full repayment was made. At that point their utility bills would again be calculated only on the measured amount of usage. For extremely low-cost improvements, but ones with a significant impact on energy bills (such as sealing ducts, adding insulation, and installing programmable thermostats), it would just be a matter of months before loans could be recouped from the savings on utility bills. Because of the larger financial resources available to them, owners of existing non-low-income homes have the greatest power to influence energy efficiency (40% of the potential), and the key ways for achieving those gains are listed in table 5.1.

The savings from most of these actions will more than pay for the cost of the retrofit for homeowners the year they are implemented, and some (like replacing an automobile with a bicycle) can create huge savings ($14,000) for residents the first year they are implemented. The most important actions householders can take suggest a few general rules for reducing energy use:

- Investing in efficiency has tremendous energy-saving potential, and it does not require sacrificing comfort or mobility.

Table 5.1 Summary of the most important energy-saving actions and their estimated payback time in the United States[1]

Activity	Percentage of household energy saved	Typical cost for action	Yearly return from energy savings[2]	Payback time
Inside the house				
Replace incandescent with CFL or LED bulbs	4	80¢–$20/bulb[3]	$53	<1 year[4]
Use energy-saving thermostat setting (e.g., A/C at 78°F)	3.4	0	$45	none
Clean clothes with warm wash, cold rinse	1.2	0	$16	none
Caulk and weatherstrip	2.5	$20–$200	$33	<1 year
Upgrade attic insulation and reduce airflow (seal ducts)	7	$300–$1,000	$93	3 years
Total	18.1	$350–$1,500	$240	1.5 years
Transportation				
Replace car with bicycle or public transportation	20–40[5]	save ≥$6,000[6]	$8,622 (for 1 vehicle)[7]	negative
Carpool to work	4.2	0	$390[8]	none
Maintain car (e.g., tire pressure)	5.1	0	$25	none
Drive smart (e.g., avoid sudden starts)	3.2	0	$16	none
Drive slowly (<60 mph)	2.4	0	$12	none
Total	34.9–54.9	–$6,000	$9,065	negative
End-of-life replacements				
Replace a 20 mpg car with a 30 mpg car	13.5	—	$165	—
Install low-rolling-resistance tires	1.5	—	$8	—
Install a 92% efficiency heating unit	2.9	—	$38	—
Install a SEER 13 or better A/C unit	2.2	—	$29	—
Replace old refrigerator with new Energy Star unit	1.9%	—	$25	—
Total	22	—	$265	—

[1]Using 2010 or 2011 dollars and gas prices of $2.88/gallon.

[2]Based on an average U.S. electricity bill of $110.55/month in 2010 (U.S. Energy Information Administration, 2011).

[3]This price range reflects LED bulbs being more expensive than CFL bulbs.

[4]Based on CFL bulbs being used.

[5]There were 1.87 vehicles per household in 2009, so replacing one vehicle would reduce almost half of the vehicular energy use, and replacing both would reduce almost the entire amount (43%).

[6]According to the U.S. Census Bureau, the average cost of a used car or similar vehicle was $8,186 in 2007. Some sellers may choose to purchase an expensive bicycle and do a poor job in selling the vehicle it replaces, so $2,000 was subtracted from the price.

[7]In 2010, the average private vehicle was driven 11,300 miles/year at a cost of 76.3¢/mile, creating an annual expense of $8,622 (*Transportation Energy Data Book: Edition 29*, 2010), based on a $2.88/gallon gas price.

[8]This value represents 15% of the operating costs, because approximately 30% of the miles driven are to and from work, and a carpool would split those costs (*Transportation Energy Data Book: Edition 29*, 2010).

- Replacing expensive items with energy-efficient alternatives at the end of their life spans saves the most energy at the lowest cost, but early replacement may create the highest cost per unit of energy saved.
- Driving cars less makes the biggest impact on energy use, because cars dictate energy use per household.

If average householders follow all the energy-saving suggestions in table 5.1, they would both cut their energy costs in half and reduce their expenditures by nearly $10,000 a year, with little, if any, cost for implementing the actions in the first place. Of course the benefits of these actions may be far higher for large, energy-inefficient houses, and these improvements are probably already completed for the few people who own Energy Star homes or other homes designed specifically for energy efficiency. New windows are conspicuously absent from table 5.1, along with solar panels and wind turbines. Because windows have such poor insulating value to begin with, even significant improvements associated with triple-glazed windows or special gases injected between the panes only create moderate energy savings, and they do so at high prices. The top-of-the-line windows cost up to $1,000 apiece, and a typical whole-house replacement would run between $7,000 and $25,000. Replacing old single-pane windows with better but pricier models would produce energy savings ranging from $100 to $450 a year, depending on local climate and energy costs, and those savings fall to $20–$100 a year if the original windows were double-pane windows (www.energystar .gov/index.cfm?c=windows_doors.pr_savemoney/). Simply put, don't replace your windows to save money unless you are willing to wait 20 years for your first returns. Fortunately, there are several new products (ranging from window shades to window tinting) that can improve the performance of existing windows at a fraction of the replacement costs.

Replacing a private motor vehicle with a bicycle or using public transportation is the single most important and profitable action a householder can take. Household transportation accounts for nearly half of a household's energy consumption, and almost all of that results from the use of private motor vehicles. The average American spends nearly as much on personal motor vehicles (15.6% of all expenditures) as on health care, entertainment, and education combined.[20] The average private motor vehicle is driven 11,300 miles per year, at a cost of 76.3 cents per mile, creating an annual expense of $8,622 for that vehicle. Unfortunately, less than half of all Americans have access to reliable public transportation, and the use of public transit actually declined (from 6.4% to 5%) between 1980 and 2009.[21] Despite the success of recent light-rail programs and the populace's overwhelming support for public transit, the political will for building such systems has been lacking. Fortunately, many householders can utilize a bicycle on their own.

Adding bicycles into the mix of household transportation has the biggest impact on the bottom line when two-wheelers replace a four-wheeled personal motor vehicle (a car, a pickup truck, an SUV, or a van). The purchase and insurance costs for the latter are higher than the cost of gasoline to run them, but riding a bicycle to work while keeping a motor vehicle for emergencies, rainy days, and occasional other trips can still rack up major energy and financial savings. Doing so would save the average householder 8%–9% of the household's overall energy use and $800 a year (double the savings of carpooling). These savings are even larger if parking costs are avoided. Despite the obvious environmental, economic, and health benefits of cycling, only 0.5% of Americans ride a bicycle to work, and that figure hasn't changed in 30 years.[22] There isn't any solid research on why people don't ride bicycles to work, but perusing public comments and articles about bicycle commuting highlights some common concerns. Table 5.2 lists those concerns, along with easy solutions for addressing them.

Perhaps the most common set of worries revolves around getting sweaty, and the obvious solution is showering. Most people shower at some point each day, anyway. Unfortunately, many workplaces do not have showers, and some people may not want to shower at work, even if facilities were available. In the latter cases, householders can use an e-bike (bicycles with small electric motors that assist the rider in pedaling or entirely power the bicycle). Most states and nations limit e-bike speeds to 20 miles per hour. E-bike prices are similar to those for traditional bicycles, with basic models ranging from $500 to $2,000, depending on the features included and the battery type. New battery technologies allow an 8-pound battery to propel an e-bike for 20 miles at 20 miles per hour (mph). E-bikes are rapidly replacing motorcycles and traditional bicycles in multiple markets, including China and portions of Europe.[23] Interestingly, e-bikes are so efficient, and traditional agriculture is so fossil-fuel dependent, that an e-bike rider may actually have a smaller carbon footprint than riders of traditional bikes, who must eat slightly more food to provide the caloric energy to power themselves.

If you are considering an e-bike, the models with lithium-based batteries are far more reliable than those with lead-acid batteries and worth the additional cost. E-bikes also address the frequent concern about how physically

Table 5.2 Common bicycle commuting concerns householders may have and solutions for them

Concern	Solution
Getting sweaty	Shower at work
	Use an e-bike and don't break out in a sweat
Getting tired	Use an e-bike and don't pedal
	Get in shape
Cycling is too slow	Admit that the average commuting speed for a motor vehicle is 31 mph
	Admit how much time you spend working to pay for your motor vehicle
	Replace your gym time with cycling time
Clothing problems	Ride your e-bike with your work clothes on
	Bring clothes to work in a vehicle once a week (or whenever you end up driving)
	Bring clothes in a bag
Bad weather	The weather is great for cycling on many days, so ride then and drive the other days
	Use rain gear
It's too dark at night	Use headlights/tail-lights and wear reflective gear
	Don't ride at night
It's dangerous	Health benefits outweigh the increased risk of an accident by 20 to 1
	Don't ride a bicycle drunk (30% of the cyclists killed in 2009 had alcohol in their system) or when motor-vehicle drivers might be drunk (12% of all deaths involved a drunk driver, and crashes were more common in the evening)
	Encourage your municipality to provide pedestrian and cycling infrastructures, based on the cost savings associated with getting motor vehicles off the roads

difficult bicycle commuting might be. With a battery, pedaling is optional. Issues associated with how to get business-appropriate clothing to work or cope with a rain shower are similarly easily addressed. Bicycles not only save money and protect the environment, but they also make roads more pleasant by reducing vehicle noise. Of course, total silence can pose a safety problem in places where roaring engines signal imminent danger. Companies using electric scooters to deliver Dominos pizza in the Netherlands addressed the problem by broadcasting engine noises and words (including "yummy" and "pizza") from speakers on their otherwise-silent delivery fleet.

The slow speed of bicycles relative to motor vehicles is often cited as another reason to avoid bicycle commuting, but the difference in speeds between cycling and commuting by motor vehicles is less than you might imagine. The average commuter speed for motor vehicles in the United States is 31 mph, compared with 10–15 mph for a bicycle.[24] At face value the speed differential equates to a 20-minute savings on a 10-mile commute. The time savings, however, is deceptive. Commuters using motor vehicles pay for that 20 minutes with their vehicle payments, insurance bills, mechanics' bills, gasoline bills, parking fees, tolls, and gym memberships. The effective speed (the distance traveled divided by time, where time includes all the minutes and hours dedicated to the mode of transport, including the time spent working to earn enough money to pay for the various costs associated with the vehicle being used) of bicycles outstrips the speed of virtually all motor vehicles.[25] For example, a family making $50,000 a year would spend 359 working hours to run a typical motor vehicle. After adding more than two months of work to that commuting time (to pay for the cost of vehicle use over that period), bicycles start looking really fast. The effective speed of four-wheeled motor vehicles increases for a user as that user's income increases. So someone making $1,000 an hour only loses between 9 and 20 hours a year to pay for their vehicle (unless it's a Ferrari). On average, the now-famous 99% (as opposed to the wealthiest 1% of the population) would travel much faster on bicycles than with cars, while wealthy commuters can actually average an effective speed of just under 30 mph in a car if they drive an average-priced car. The effective speed for bicycles is even higher if you replace exercise time in gyms with time on a bicycle, or recognize that cyclists are healthier and spend less time and money on medical care than other commuters.[26]

Danger probably ranks as the most important factor in concerns about riding bicycles to work. Although less than 1% of all commuting is done by bicycle, 2% of traffic fatalities are bicyclists,[27] leading to a conclusion that bicycling is more dangerous than driving a motor vehicle. That said, bicycling is still extremely safe, given the hundreds of millions of bicycle trips that are taken each year, and the fact that more drivers suffer traffic fatalities than bicyclists. Further, nearly a third of the bicyclists who have been killed in accidents had alcohol in their system, and half were killed at night. These facts suggest that the risks associated with bicycling are largely under the control of the user. Not traveling without lights at night, not drinking and cycling, and not breaking traffic laws make bicycling far safer than it otherwise appears. For the more altruistic among us, small increases in the number of bicycle commuters create drastic declines in accidents by forcing drivers to recognize that bicycles are on the roads.[28] For instance, in California the risk of being hit by a motor vehicle declines nearly tenfold in communities where more than 2% of the residents commute on bikes, relative to places where less than 1% commute by bicycle. Furthermore, some research suggests that the overall risk of mortality among bicycle commuters is lower than that of the general population, because the health benefits of bicycling outweighing its safety costs.[29] You might want to be extra careful, however, in Arizona, Delaware, Florida, Hawaii, and Wyoming, where annual bicycle-related deaths are more than twice the national average of 2.05 bicyclists per million.[30]

Making Energy

Sustainable sources of household energy may conjure up images of solar panels and windmills, but these highly visible symbols of environmentally conscious behavior are typically the most expensive, least effective, and last options householders should consider. Currently, the cost of installing wind power is negatively related to the size of the wind turbine. So the cost of electricity produced by small household turbines is often even more expensive than photovoltaic solar energy (the kind produced by solar panels). Huge turbines on towers 50 meters (164 feet) tall can produce energy economically in areas where the wind regularly blows at more than 10 miles per hour, but, for obvious reasons, people living in urban areas cannot install economically sized turbines on their lots, and most people live in urban areas. In 2010, over half the world's population lived in urban areas, as did over 75% of the people in developed nations. Householders have more options with photovoltaic solar cells, because panels are the most expensive piece of the systems, and they produce power in direct proportion to the area they cover. Thus the savings associated with economies of scale for massive photovoltaic systems are smaller than similar savings for wind-based systems. Moreover, having small-scale photovoltaic systems installed on homes can cut energy costs by reducing the roughly 7% of energy lost during the long-distance transmission of electrical power.

Household generation of solar power, however, is still far more expensive than making changes to reduce one's energy usage. For example, offsetting half of the monthly 1,000 kilowatt-hours of electricity used by an average house in the United States would require a 5-kilowatt solar- or wind-powered system. The costs of these systems vary drastically, based on the quality of the materials and the degree to which homeowners would do the installation themselves, but the unsubsidized cost for a 5-kilowatt solar- or wind-powered system was between $25,000 and $50,000 in 2010. Without subsidies, the payback period (how long a person would need to wait before the cost savings associated with their investment equaled the initial investment) for such a system in areas where solar power is readily available (e.g., in the Southwest, and for homes not shaded by trees) would range from 30 to 80 years. It is true that huge subsidies for solar power can drastically reduce the payback period, but someone is paying for those subsidies, even if the homeowner is not. Most householders could cut their electricity use by over 30% by making minor behavioral adjustments (e.g., changing their thermostat settings), investing no more than $1,000 in sealing ducts (thereby reducing air leaks) in their homes, and adding insulation (table 5.1). Thus creating the energy required (by installing solar- or wind-powered systems), instead of saving it by reducing the household's energy needs, could cost 50 times more money. Although household energy production is an exciting topic, the economic rewards of energy efficiency are far more immediate.

Saving Water

Conserving water is another key practice householders can engage in to make their homes more sustainable. Like energy conservation, most water-conservation activities not only help the environment, but they also save money. Reducing household water use has obvious benefits: keeping the water in streams and rivers for fish and other aquatic animals, avoiding the costly construction of reservoirs, and preventing catastrophic drops in aquifers.[31] The average family of four in the United States uses 400 gallons of water per day, or 146,000 gallons per year, and 70% of that is used indoors. Delivering and treating water is energy intensive, so conserving water reduces energy use and saves money. The Environmental Protection Agency (EPA) estimates that letting a faucet run for five minutes uses as much electricity as turning on a 60-watt lightbulb for 14 hours.

The EPA's WaterSense partnership program is rapidly bringing householders the same reliable labeling needed for water conservation as the Energy Star program brought for major appliances. The WaterSense label applies to many water accessories, but toilets are the main culprits (more than 30%) behind household water consumption. According to the EPA, replacing older toilets with WaterSense models that use less than 1.3 gallons per flush will save the average family of four about $90 per year in lower water bills, or $2,000 over the life of the toilets, and cut home water use by 18%. Although many utilities offer rebates for upgrading toilets (and other fixtures), the

WaterSense models have the same starting price as other models (e.g., less than $100 for a toilet) and can be paid for entirely by water savings in approximately one year, without any rebate. Low-flow shower heads can save 12% of a household's water use, or $60 a year, and also pay for themselves in less than a year. End-of-life replacements of dishwashers and clothes washers with Energy Star models can save 5% on household water use.

Other water-conservation measures (including rain barrels and composting toilets) have gained popularity in recent years, but they may still have limited applicability. With average water costs ranging from $2 to $10 per 1,000 gallons, a 60-gallon rain barrel priced at $100, and used as an alternative to a municipal water source, would need to be filled roughly 167 times to pay for itself. Only a few serious gardeners in areas where rainfall is relatively evenly distributed over time could achieve reasonable payback times for rain barrels. Of course, if residents like the barrels for aesthetic reasons, or because they provide a water source containing fewer chemicals, or because they help reduce polluted runoff into local streams, payback may not be an issue.

Similarly, composting toilets can operate without water, but most systems cost between $1,500 and $6,000.[32] Assuming $150 is saved each year in water costs, it would require 10–40 years for a householder to pay for this system with water savings. Further, composting systems require regular care, and they create health, odor, and aesthetic problems if their users are not committed to maintaining the system. These systems, however, become much more attractive when viewed as alternatives to installing expensive septic systems, or as a means to manage human waste in environments where traditional sewer and septic systems are not possible (e.g., solid rock, hardpan, or wetlands), or where they are not adequate (in urban areas of many developing countries). In developing countries, where expensive composting systems are not an option, householders can construct their own composters out of relatively common materials. These do-it-yourself composting toilets may have more odor problems than more expensive systems, but they are sanitary, safe, and much better than open sewers and fouled waterways.

Saving Nature

Using native plants for landscaping is another way to reduce water consumption, and it can provide important benefits for wildlife and local ecosystems. The physical footprint of cities makes home landscaping a major factor in the decreasing quality of water and soil, the loss of biodiversity, and climate change. Urbanization can contribute to sustainability when cities are densely populated, but urbanization in many developed nations has recently been characterized by sprawling suburban neighborhoods.[33] This trend means that householders own and make management decisions for ever-greater portions of the urban land area, and their decisions shape the sustainability of cities.[34] Householders also may influence the vegetative cover on public lands near their homes.[35] Turfgrasses are often desired by householders (perhaps in part because they are heavily advertised), and in 2005, they grew on more than 16,380,000 hectares (49,475,699 acres) in the United States, an area three times larger than that dedicated to corn.[36] That turfgrass area is expanding annually, with 23% of new urban lands (675,000 hectares [1,667,955 acres] per year) dedicated to turfgrass lawns.[37]

The production of turfgrass alters biogeochemical cycling and the global carbon cycle.[38] The maintenance for this type of landscape contributes to environmental degradation through the use of fertilizers, pesticides, and herbicides, which damage water and soil quality; the increased use of lawn mowers adds to the growing amount of carbon dioxide emissions, which are linked to climate change; and irrigation and other forms of watering draw down limited water supplies.[39] In the United States, 30% of all household water is devoted to outdoor uses, and the bulk of this water irrigates turfgrass. In desert climates, the quantity of water a homeowner uses in one year to maintain turfgrass can be greater than the amount used for all other purposes. Finally, turfgrass-dominated landscapes are sterile in terms of wildlife habitat, because they lack the necessary vertical and horizontal vegetative structure, as well as the native plant species that wildlife require for food, cover (protection and places to hide), and reproduction.[40] Alternatives to turfgrass-dominated landscape designs can promote a number of ecosystem functions simultaneously, including moderating extremes in local temperature, sequestering carbon, reducing air and water pollution, and providing habitats for birds and urban wildlife.[41]

Fortunately, householders have many alternatives to turfgrass, and they appear to prefer some of those substitutes. Native-plant gardens are an example of an ecologically friendly landscape design that may reduce the use of chemicals, energy, and water.[42] According to the EPA, the typical suburban lawn consumes 10,000 gallons of culinary water for irrigation each year. Using native plants can virtually eliminate this water use, decrease soil erosion, reduce maintenance costs, and create wildlife habitat. Because native plants are adapted to the local climate, they will save homeowners time and money by requiring no fertilizer, minimal or no watering, and limited control of competing species (e.g., invasive plants and weeds). In many regions (the Northeast, the mid-Atlantic states, the Midwest, and the Pacific Northwest) stopping turfgrass maintenance and then waiting will yield a naturalized landscape of native plants that will recolonize the area. This process typically takes two to five years. Ceasing to water and waiting for other plants to take over turfgrass areas is generally not a successful option in other, drier areas (such as the West, the Southwest, and the Rocky Mountain region). In these regions, xeriscaping (a landscape designed to withstand drought by using plants with low water needs) can save householders time and money. Xeriscaping also requires less maintenance, less fertilizer, and fewer pesticides than turfgrass.

Despite the economic and time-saving advantages of native plants and xeriscaping, turfgrass continues to dominate landscaping, both for homes and recreational areas. A preference for turfgrass is the obvious explanation, but that supposed preference doesn't hold up under scrutiny. Rather, research suggests that residents' preferences are swayed toward the landscaping designs prevalent in their neighborhoods (peer pressure), and they worry about what their neighbors prefer. A survey conducted in Raleigh, North Carolina, found that residents preferred a 50% native-plant garden design over a 100% turfgrass design, but mistakenly believed that their neighbors liked the pure turfgrass best.[43] Native landscaping gains acceptance with cues to care (landscaping elements that demonstrate a homeowner is controlling a landscape), which are typified by colorful flowers, borders, or canopy trees. They suggest order, the influence of labor, and a respect for nature.[44] Landscapers and householders need to know that turfgrass isn't as exciting to residents as it once may have been.

Householders can overcome neighborhood norms against the potentially "messy" look of native landscaping by using the National Wildlife Federation's Certified Wildlife Habitat program. The sign provided to participants shows all the neighbors that an innovative homeowner is certifiably creating wildlife habitat rather than merely neglecting turfgrass (fig. 5.1). The program requires householders to establish yards that provide wildlife with food, water, cover (places to hide), and places to raise their young. The program also encourages replacing exotic (non-native) plant species with native species, eliminating chemical pesticides and herbicides, composting, capturing rainwater, using soaker hoses, minimizing lawn coverage, and xeriscaping. This program is designed to have broad appeal, and to be achievable by all

Figure 5.1 The National Wildlife Federation's Certified Wildlife Habitat sign posted at a Florida home.

householders. It's relatively easy for any landscape other than one with 100% turfgrass to qualify as wildlife habitat, particularly if a homeowner installs a birdhouse or a wildlife feeder. This raises one issue, though. Providing supplemental food to wildlife using feeders typically increases the risks of both disease and predation for the wildlife. This apparent flaw in the certification program can be negligible, however, and it does reflect the fact that wildlife feeding, particularly bird feeding, is enormously popular. The program could alienate many potential participants by denigrating the role of feeders.

The best landscaping decisions for wildlife may actually be the easiest, cheapest, and most beneficial for the environment: facilitating or simply allowing the regeneration of native plants. These plants rarely (if ever) need water or fertilizer, and they provide the habitat local wildlife have adapted to. Native plants can supply all the food resources that wildlife need. Cooperative Extension Service programs in most states provide free guides indicating ways to obtain native plants, the various conditions they thrive in, and whether they provide food and cover for wildlife. Similarly, there are publically available guides for creating backyard habitats for birds, reptiles and amphibians, and other wildlife that are specific to particular regions, but they have some commonalties: (1) provide water, (2) provide food (preferably through native plants), and (3) provide cover (e.g., brush or rock piles).[45]

Creating wildlife habitat, however, can be a tricky business. Planting the host vegetation required by many butterfly species can create habitat and attract the butterflies to urban areas.[46] Doing so promotes butterfly conservation, unless a municipality sprays for mosquitoes, which also kills entire populations of butterflies attracted to the host plants. Similarly, over half of all homeowners feed birds, with about 50% using bird feeders.[47] Unless these feeders are regularly cleaned—and few are—they spread disease. If the feeders are not carefully monitored and strategically placed, they attract predators (e.g., raptors and outdoor domestic cats). Further, feeding wildlife can create ecological sinks, where wildlife shift their distributions into urban areas to be near food sources, resulting in lower fitness (the ability to survive and reproduce) in that urban environment. The relationship may seem counterintuitive, but increased wildlife mortality caused by exposure to toxins (e.g., pesticides), diseases (e.g., those transmitted at feeders and birdbaths), and multiple predators (e.g., other wildlife, pets, and vehicles) often offsets any benefits associated with easy access to large volumes of energy-rich food.

Household pets are another key impact households have on the environment. Escaped pets, ranging from pythons and iguanas to cats and dogs, wreak havoc on wildlife populations. For example, around 600 million domestic cats exist globally, and 50–150 million of them roam freely in North America alone.[48] These free-roaming house cats kill billions of wild animals every year and responsible pet ownership requires keeping one's cats indoors.[49] Recent research suggests one out of three house cats are successful wildlife hunters, and that these cats kill an average of 2.1 animals every week. Cat owners typically do not notice the extent of this depredation, because cats bring less than one in every four kills home.[50] In many cases, moving free-ranging domestic cats inside the house is vital to prevent the extinction of native species.[51]

Barriers to Sustainable Households

The difficulty in promoting sustainable decisions about households stems from two primary types of barriers: structural and behavioral.[52] Structural barriers occur when physical (including time), legal, or economic environments stymie sustainable decisions. The physical absence of quick and reliable public transportation presents a notable structural barrier to energy efficiency. When that structural barrier is overcome, the adoption of public transportation on a massive scale occurs in both developed and less-developed nations.[53] This rapid acceptance and use of public transit has taken place even in areas like Utah, where public transit has generally been opposed. A light-rail system in the Salt Lake City area probably was approved only as bait to attract the 2002 Winter Olympics, but once public transit was available, use skyrocketed and the system was expanded to accommodate the rising demand from new users. In the United States, overcoming barriers to public transportation could cut household energy use by almost 40%.[54]

Transaction barriers, including the substantial amount of time needed to sort through various energy-efficiency investments, provide another structural barrier for residents. Even if homeowners have sufficient financial capital, they must sift through all the advertising claims about the energy-saving benefits of aluminum attic blankets and metallic shingles before finding out that installing programmable thermostats and sealing ducts are much more effective measures in terms of reducing energy use, and they are far cheaper. As energy becomes progressively more expensive, transaction barriers will naturally decrease. Further, a growing body of information is beginning to carefully highlight the most cost- and energy-efficient strategies for reducing household energy use (table 5.1). The EPA's governmental labeling programs, such as Energy Star, WaterSense, and Fuel Economy and Environment stickers on new motor vehicles, have given consumers an easy way to identify appliances, houses, plumbing fixtures, and vehicles that save money and benefit the environment. Third-party testing has shown that these products are as reliable and effective as their counterparts. Care must be taken, though, to avoid "gold plating" in association with these labeling programs, as energy efficiency has a tendency to be bundled with unrelated premium features in vehicles and appliances. The WaterSense label seems to have avoided this problem, as most stores stocking plumbing items sell WaterSense toilets and fixtures at prices similar to their water-hogging counterparts.

Legal structural barriers occur when certain decisions are relegated to landlords rather than tenants, or when a homeowner must follow rules enacted by homeowners' associations, municipalities, counties, or states. In these contexts, legal arrangements and regulations can prevent the installation of insulation, light-colored roofing, or new windows. About 8% of the energy-efficiency potential inside homes is impeded by the structural barriers associated with rentals.[55] Tenants and landlords currently do not have a way to split the financial benefits associated with energy efficiency, and addressing this barrier requires innovation. If landlords were required to pay for the energy used in their rental units, they would have an incentive to invest in energy efficiency. This sector should be particularly easy to motivate, because, as business professionals, landlords might have a more reasonable expectation for payback periods on their energy-saving investments than the average homeowner, who expects almost immediate returns. This type of legal structural improvement could create behavioral problems if tenants lose all accountability for how they use energy, but charging a fee if a tenant's energy consumption surpasses a baseline amount would address such problems.

The economics of market failures prevent a host of energy-efficiency improvements. Ownership-transfer problems occur when owners cannot capture financial benefits from their improvements before they sell a property, and they are unable to fully recoup the benefits of energy-efficient upgrades when they do sell it. Ownership transfer creates an important economic structural barrier that impedes 40% of potential retrofits that would aid energy efficiency. The 17-year average lifespan of energy-efficiency measures is far longer than the 7-year average payback period, but 40% of the homeowners in the United States will leave their dwellings in less than seven years. This barrier can be addressed by ensuring that homeowners recover the value of their energy improvements when they sell a home, perhaps by providing a cost-of-ownership label when homes are sold or rented to new tenants. Cost of ownership, as part of the monthly sums that new owners or renters expect to pay, would include the average utility bills per square foot for the home. This approach would also encourage the most important (yet relatively cheap) conservation improvements, because they would have the greatest impact on the advertised cost of ownership. Improved labeling of energy-efficiency upgrades, and legislation requiring such upgrades during sales or when a unit is rented to a new tenant, would also help homeowners recover the costs of their energy-efficiency upgrades and water-conservation measures.[56] In the case of new houses, cost-of-ownership estimates could still include projected energy and water bills, but these would be based on construction practices. Such programs could be particularly effective, because $1,500 worth of energy-efficiency additions to a house would increase the monthly mortgage payment by $10 but decrease the monthly utility bills by $30–$40. It's a clear choice for consumers if they have this information. The Energy Star Home program has begun to supply such data for those purchasing new homes. Used homes, however, do not fit into the program well, although their energy-saving potential is four times greater, so labels based on past energy and water use are crucial in the used-home sector.

The current inability of markets to attach a price to the long-term environmental impacts of decisions provides a classic structural barrier to sustainability. For instance, renewable energy seems less appealing because the cost of electricity produced using fossil fuels does not factor in the costs of fighting its eventual consequences, such as rising sea levels. Similarly, homes in city centers seem less appealing than those in suburbs, because a householder abandoning the city for suburbia is not forced to pay for the disproportionately high long-term costs of expanding and maintaining roads, sewers, and schools to support farther-flung homes. This problem is exacerbated when people who leave decaying neighborhoods escape paying for the social services (e.g., schools) needed in the neighborhoods they abandon. Market failures occur when products desired by consumers are not manufactured in sufficient quantities or are produced with "gold plating."[57] The gold-plating problem can be seen in the emerging market for electric cars; none come without accessories typically provided only in premium packages for other models (e.g., touchscreen onboard computers). Public transportation typifies the insufficient quantity problem. Reliable public transit is not available in most locales, so consumers do not have the option to use it.

Behavioral barriers explain why a resident who is structurally able to capture benefits from making sustainable decisions about energy use still chooses not to make them. These barriers may be the most important, because overcoming them can achieve energy savings at incredibly low costs.[58] Some choices that are prompted by custom and habit (e.g., driving instead of bicycling) have by far the largest potential to reduce energy consumption. Nationwide research has demonstrated that sending householders a mailing comparing their household energy use with that of similar homes created a 2% decrease in their energy consumption. If this type of program were to be adopted nationwide, it would save $3 billion, with an expenditure of less than $1 billion (using 2010 prices). The postage for and coordination of this effort cost 2.5 cents per kilowatt-hour saved, while producing the electricity was four times as expensive (at an average cost of 10 cents per kilowatt-hour). Nudging people to act can have a more positive influence on behavior than imposing large increases in energy prices.[59] Conservation efforts that rely on psychology (internally generated rationales) instead of technology (externally applied levers) to change behavior have additional benefits: they save money by getting people to reduce their energy usage, rather than trying to find a lower-cost way of generating the same amount of energy that consumers have habitually used. For instance, designing an advertising campaign to persuade individuals to decrease the amount of electricity they consume, offsetting a metric ton of CO_2 being created by conventional power-generation sources, would result in a net programmatic gain of $165 (after the cost of electricity was considered), whereas offsetting a metric ton of CO_2 through the use of a different technology to produce that energy (wind turbines) would cost $20.

Behavioral barriers typically arise from one of four causes: uncertainty, a lack of knowledge, misinformation, and norms (habits). The inability to trade the use of automobiles for public transit and bicycles is a prime example of a behavioral barrier. Anyone anywhere in the world who drives a car (or a pickup truck, an SUV, or a van) is structurally able to replace that vehicle with a bicycle, because a motor vehicle in running condition can be sold for more money than it costs to acquire a high-quality bicycle, but few do so, because they are habituated to driving and are misinformed about the costs, the degree of independence, and the speed (the amount of time saved) associated with their personal vehicles. Drivers underestimate the true overall costs of their motor vehicles (expenses beyond the amount of gasoline they use) by at least 60%.[60] They overestimate their independence, because they do not associate taxes, the cost of infrastructures (e.g., interstate highways), and pollution with their choice to drive. They also overestimate speed, because they do not recognize that the many hours they spend working to pay for and maintain their vehicles is a part of "speed."

Similarly, norms prevent people from modifying the impact their households have on the environment. Although householders are willing to expend money on visible improvements with little hope of monetary payback (e.g., kitchen remodeling, high-end windows, and top-of-the-line appliances), they expect complete payback for less-visible environmental improvements (e.g., insulation) in 2.5 years.[61] This equates to a 40% annual discount rate (the return on their investment), about 10 times higher than rates average investors accept for stocks and bonds. Most research suggests that householders want to reduce their energy use and the amount of waste they produce, but they are either unsure about the most effective steps to take or misinformed about them.[62]

Put Your Money Where Your Mouth Is

When we started this book in 2009, we realized that we would need to personally put our advice into practice. Hypocrisy has been an Achilles heel of environmentalism for decades. Countless authors advocate not having chil-dren (or not having more than one or two) after producing multiple progeny, write about the virtues of urban high-rises while living in suburbia, argue for multigenerational households while living by themselves, and jet-set around the world to give speeches about the dangers of global warming. If house-holds provide a better approach for addressing sustainability than concerns about overpopulation, because the impacts of houses on the environment can be addressed immediately and in ways that benefit everyone involved (including householders, their communities, and society in general), then why wouldn't we make these changes?

As it turned out, one of our houses (Nils Peterson's)[63] was a perfect ex-ample when it came to testing the recommendations made in this chapter about how to defuse the housing bomb with your own dwelling. This home measures 2,154 square feet, just 15 square feet shy of the 2,169-square-foot median in the United States.[64] In other words, about half of the homes in this country are larger and half are smaller. Further, this home was typical in that the heating and air-conditioning systems and the ductwork were improperly installed, there were air leaks everywhere, and all the plumbing fixtures were water-guzzling models installed in the late 1980s, before regulations requir-ing water-efficient toilets were enacted. The average commute for household members was 11.7 miles one way, just short of the 12.2-mile national average.[65]

Our experiment was guided by the recommendations in table 5.1. By fol-lowing them, transportation costs dropped by $1,245 a year; electricity use in the house went down by 49% (from a monthly average of 745 kilowatt-hours to 380 kilowatt-hours), saving $465 annually (energy costs in North Carolina are among the least expensive in the United States); water use fell by 20% (from 3,100 gallons per month to 2,500 gallons per month), saving $85 per year; and natural gas use decreased by 28% (from a monthly average of 55.3 therms to 39.9 therms), saving $222 a year. The cost reductions from these improvements are now saving us a total of $2,017 a year. The average house-holder would save far more from these same changes, however, because the norm is at least a 30% greater use of electricity, gas, and water than ours, even before we started making changes. Our experiment also resulted in a lesser amount of savings than most could expect, because electricity, water, and natural gas prices in our area are well below national averages, and $3-per-gallon gasoline prices are now a thing of the past.

We spent $7,350 on all of the changes we made, so it will take 3.5 years cover our outlay for them with the savings they provide. After that, these improvements will result in savings of at least $2,000 annually, since utilities will certainly become more expensive, and gasoline already has. Most of our expenditures were associated with replacing a failed heating, ventilation, and air-conditioning (HVAC) unit ($4,000). Buying an energy-efficient model was cheaper than an alternative, less-efficient one, thanks to rebates. Also, $1,950 was used for a sealed and air-conditioned crawl space that we paid contractors to install, mainly to get extra storage space. We could have achieved similar efficiency gains by sealing all the gaps from the crawl space to the first floor, which would have cost just $100. The more efficient HVAC unit potentially makes some contributions to energy savings (2% for cooling and 2.9% for heating), but these figures are relatively modest compared with the 30%–50% decline in our overall energy consumption. If our outlays for the new HVAC unit and extra storage space are excluded, the remaining changes cost $1,400. After only six months, they repaid our initial investment and started saving us $2,000 a year.

The biggest change was in transportation, which entailed converting a 15.4-mile commute from a car trip to an electric bicycle trip. We purchased a Prodeco Phantom X electric bicycle for $1,199.99. The e-bike has a 500-watt motor that propels a rider at 17 mph without pedaling (we tested this with a cycle computer), and a lithium-ion battery that runs for 18 miles on a single charge. The battery is rated to last for more than six years with greater than 80% capacity, and it will last indefinitely afterward, but with diminishing power. We racked up about 3,700 miles during the trial year and saved almost $1,250 (more than covering the price of the new toy): $672 worth of gasoline (168 gallons) that would have been pumped into my 22-miles-per-gallon car, $400 on a forgone parking pass, and $175 on maintenance and tires, the latter figure coming from the American Automobile Association's 2011 estimates of driving costs. Next year the e-bike will double our savings, and be a far better investment than any of our retirement accounts. The reduced expenditure on gasoline would be even more exciting (at $1,400 a year) if we started with a clunker that got 12 miles per gallon. We parked the e-bike for free right in front of the office door. (The only other way for a professor to park for free at his or her office is to win a Nobel Prize.) Riding the e-bike is much like commuting on a motorcycle, but without paying for insurance and gasoline, or making loved ones fear for the rider's life. E-bike riders also have the option, which was used on occasion, of throwing it in the back of a colleague's car for a ride home. The e-bike was the best transportation choice, because it was three times faster than public transportation (30 minutes versus 1.5 hours on buses), and the commute was relatively long (over 14 miles round trip).

Making energy-use reductions inside the house was more complicated. Replacing lightbulbs ended up being more expensive than we planned, because we bought seven light-emitting diode (LED) bulbs in addition to a handful of compact fluorescent (CFL) bulbs. The CFL bulbs cost less than 50 cents each, as they were subsidized by the local power company. The LED bulbs were $10, but they could be dimmed. We wanted to try them out, because they are supposedly the wave of the future. The two types of bulbs have expected life spans of more than five years (greater than 15 years for the LED bulbs), so never having to change lightbulbs has been nice. We also installed a programmable thermostat (for $25). The installation was easy and safer than changing a lightbulb, because the attachment wires for thermostats are low voltage. We cheated by creating a custom setting instead of accepting the pre-programmed Energy Star settings, because keeping the house at 78° F at night during hot weather was too warm for us. The programmable thermostat also ensured that the temperature settings were adjusted at the appropriate time every day: turning on just before we awoke, changing when we left for work, coming on again when we were back home, and adjusting at bedtime. Like many of the other changes, the programmable thermostat increased the comfort of our home as well as saving us money.

We got expert help in our efforts to seal the house. The contractors sealed off the crawl space and ducts, and we took care of the living spaces and attic. Sealing one's house is critical, because most homes have a chimney effect, where cooler air is pulled into gaps and cracks in the lower portions of the house and warmer air is sucked out of gaps and cracks in the higher parts, primarily the attic. Even tiny cracks in well-built homes, such as those in the backs of power outlets, can add up to the equivalent of leaving a standard window wide open. Similarly, many houses (like ours) have unsealed ductwork. The ductwork is supposed to be glued to registers and to the main HVAC system, using a substance called mastic that prevents air leaks. In many houses, however, the ducts are simply attached with screws or plastic zip ties, leaving hundreds of pencil-sized holes for conditioned air to flow into attics and crawl spaces. The contractor sealed the ductwork for $190, and we received the entire amount back in the form of a rebate from our local energy company. We could have sealed the holes between the crawl space and our living space with about $50 worth of caulk and expanding foam, but we chose to hire a local firm to "encapsulate" the crawl space for $1,950. Sealing the entire crawl space involved coating the floor with a thick plastic sheet, filling gaps and cracks with caulk or expanding foam, insulating the walls, and removing the old insulation from the underside of the home's floor. A small vent in the existing ductwork was installed to air-condition and pressurize the space. Although professionals who seal and condition crawl spaces claim resulting energy savings of 10%–30%, many of those benefits could be achieved by simply sealing the holes in the home's floor (to prevent a chimney effect) and properly insulating the floor. We, however, wanted the added benefits of an air-conditioned storage area and the elimination of mold, since allergy sufferers live in our home. Obtaining a conditioned storage area the same size as the house, while reducing allergies and saving even a small amount in energy bills, seemed like a win-win scenario.

We sealed the living spaces and attic with $50 worth of materials: weatherstripping, a caulk gun and caulk, and three cans of expanding foam. Sealing air gaps in the windows and doors was the easiest part. Sealing electrical outlets and light fixtures was trickier. For the outlets, we had to either poke around live wires with the caulk gun or turn off breakers for each section of the house where we were sealing outlets. Sealing outlets was rewarding, though, because before the work was begun, we could actually feel air blowing through them. Sealing the fixtures involved teetering around on rafters in the attic and rooting through insulation to find the fixture boxes before spraying around them with fire-retardant expanding foam (unpleasant work, to say the least). We had an adequate volume of attic insulation, so supplementation was not needed. Although we did not want to make any end-of-life replacements, our combined heating and air-conditioning unit died in 2010. The 30% (maximum $1,500) federal residential-energy tax credits made paying a few hundred extra dollars for an energy-efficient unit a profitable decision, even before considering potential energy savings. Those tax credits unfortunately expired in 2011, during the quagmire of partisan politics that engulfed the nation at that time. Other critical, and inexpensive, federal programs essential for making housing sustainable (such as funding to build safe pedestrian routes for children walking to school) were also placed on the chopping block (and ultimately reduced) during the same period.

We addressed water savings by replacing three old 3.5-gallon-per-flush toilets with newer models. One new toilet was the standard 1.6-gallon-per-flush model mandated since the mid-1990s and cost $150. Before we installed the other two, we discovered the WaterSense program, which certifies 1.28-gallon-per-flush toilets as both effective and efficient. Certified toilets can flush a kilogram (2.2 pounds) of soybean paste (used as a substitute for human waste when testing toilets) cleanly every time. Unless you share the toilet with a grizzly bear, the new models work exceptionally well for their intended purpose. Our two WaterSense toilets were paid for by $100 rebates from our municipal water provider. We learned about the rebates on the WaterSense web page (www.epa.gov/watersense/), which has links to most cities that use rebates to encourage water conservation. In our case, replacing three toilets will take about 1.5 years to pay for their cost through lower water bills, and they will then generate $85 a year in savings.

Nils Peterson had never been particularly motivated to create a beautiful turfgrass lawn, so he rarely mowed, never watered, and never fertilized the landscaping for his house prior to this experiment. Accordingly, there were not many savings from resources that had been used in landscaping, but there was lots of room to improve appearances. The sad-looking patches of grass (tall fescue) mixed with clover were retained in a small portion of the front yard. We killed the English ivy (an exotic species) that covered the backyard and allowed native plants to recolonize the area. We also planted some native food-producing plants (blackberries) in the front yard, between the relatively homely lawn and the street. So far the neighbors haven't complained, probably because the shallow clay soil in the neighborhood makes the lawns of even the most dedicated turf-tenders fall short of the ideal.

Conclusion

The suggestions presented in this chapter can help householders, their communities, and even their nations reduce their ecological footprints by 30%– 50%, and they can create major steps toward defusing the housing bomb. These adjustments do not require political change (unlike alliances among nations) or large amounts of discretionary income. They also do not ask householders to sacrifice mobility, comfort, or a particular lifestyle. Rather, these changes will enhance the quality of life for residents in a home. Riding a bicycle or taking public transit improves one's physical health,[66] and having a properly sealed and ventilated home reduces respiratory illnesses. Finally, these suggestions significantly decrease expenses for those who try them. The beauty is that they can be carried out by anyone who lives in a home. There simply is no excuse for not making our homes more sustainable, particularly now that such investments have far higher rates of return than any other investments an average householder can make. The priceless benefits to society, to our children, and to the environment do not even need to be factored into the mix for investments in household sustainability to be a great idea.

CHAPTER SIX

Individual and Local Strategies for Defusing the Housing Bomb

This chapter explores how policies operating at household, neighborhood, and city scales may help to defuse the housing bomb by curtailing sprawl (and its associated geographic segregation of people from work and from each other) and promoting viable and healthy transportation alternatives to motor vehicles (automobiles, pickup trucks, SUVs, vans, and the like). Some readers may associate the material in this chapter with the term social engineering—a label used by established elites to criticize any efforts to change the de facto forms of social engineering already operating in society—and that label is absolutely correct.[1] The social engineering that has been operating for the past century has produced the housing bomb by promoting poorly constructed, energy-inefficient, stand-alone homes; giving various forms of large subsidies to suburban development; making high-density development (e.g., clustered homes) outside city cores illegal; banning apartment complexes in certain areas; and subsidizing automotive travel by using tax dollars or tax breaks to support the oil industry and automobile-related infrastructures. This form of social engineering has perpetuated segregation for decades, contributed to an obesity epidemic, added to declines in social capital (the value of community relationships), driven many wildlife species to extinction, created pollution problems, strained limited water supplies, and made the U.S. economy the most vulnerable to declining oil supplies among all of the developed nations.[2]

At one point, the social engineering that built the housing bomb appeared to be just what was needed to improve people's quality of life and their standard of living. But much has changed since 1945, and we suggest that it's time for a different approach to household dynamics. So, yes, this chapter advocates social engineering, but not a unified version, where everyone is compelled to goose step into a utopian future. Rather, we are urging householders and communities to work together to select the policy interventions that are most effective where they live. What works well locally may not work well nationally or internationally, but promoting an alignment between local action and larger-scale outcomes should ultimately be the responsibility of state and national governments (see chapter 7). In order to help defuse the housing bomb, policy options will need to make existing housing more livable and economically sustainable, appeal to the rational self-interest of householders, and contribute to desirable community development.

Neighborhood Solutions

Neighborhood norms can be directed toward defusing the housing bomb. Daniel McGinn used the term house lust as his label for the house-addiction phenomenon.[3] The extent of this addiction is reflected in a host of reality television shows about house hunting, house flipping (purchasing a home and quickly reselling it for a profit), remodeling, and extreme remodeling. Technology has allowed householders to look up the prices paid for houses in their neighborhood and use realtors' websites to peruse the insides of homes for sale. Many homeowners are already searching for either a bigger house or exciting new upgrades the day after they purchase a new home.[4] Householders have become obsessed with housing values, and rightly so, since they represent the largest chunk of accumulated wealth for most. Current policies have shunted people into larger and larger mortgages on homes that eventually become most owners' only significant investment (although not a very wise one, from an economic perspective).

When the most recent housing bubble burst, the obsession with homes proved extremely painful for house addicts, particularly for those low- and moderate-income householders who lost their homes. In this chapter, we suggest that it is time to try something different, instead of enabling the continuation of these addictive behaviors with policies such as the Housing and Economic Recovery Act of 2008. Although rock bottom is different for every individual addict, one commonality is the sense of being completely shattered—emotionally, financially, mentally, physically, and socially. Perhaps the subprime mortgage crisis can provide sufficient motivation for modern society to see beyond their granite countertops and elegant walk-in showers.

The first key to changing household policy/governance, especially at the level of the individual householder, is education. Education is often used as a synonym for persuading people to agree with the person providing the education, but in this case we believe that the public already agrees in part with our proposals, and that they will demand new kinds of housing once they are presented with accurate information about the full costs of their current housing approach. In some contexts, householders already desire sustainable choices, but they fail to act on their preferences because they erroneously believe that their neighbors would oppose the changes, or because they overestimate the difficulty of making such changes. For example, we found that residents of Raleigh, North Carolina (the third-fastest-growing, sprawling metropolitan region in the United States, following Greensboro, North Carolina, and Riverside, California)[5] preferred native-plant landscaping over turfgrass lawns, but they believed that their neighbors would oppose any efforts to remove turfgrass.[6] Their neighbors also preferred native plants, yet they, too, believed others would object to the use of native landscaping. Because turfgrass was a dominant part of the yards in our study area, descriptive norms (standards indicated by what people do)[7] strongly suggested that the residents preferred turfgrass. These norms thus create social pressure to behave in certain ways.[8] The same phenomenon occurs with the houses situated in the landscaping. Now that the concept of walkability has been brought into the real estate conversation, the values of homes in walkable neighborhoods have begun to outpace the prices of those in spread-out suburbs built for motor vehicles.[9] In traditional suburban centers like Dallas/Fort Worth, housing associated with public transit has spiked in value.[10] Although these shifts in economic values represent the preferences of potential home buyers, sprawling suburbs and houses built around garages still dominate the landscape, and they create erroneous descriptive norms about what people want. Overcoming this problem requires education, not only of homeowners, but of landscapers, builders, and real estate agents.

Moreover, most householders simply do not understand the full costs of home ownership or the benefits associated with more sustainable houses. They underestimate the costs of commuting in their personal vehicles by 60% or more, and they rarely know (or account for) the expenses from monthly utility bills when calculating how much it costs to live in their home, including not realizing how large those expenses can be when looked at on an annual rather than a monthly basis. These educational problems are compounded by an inadequate understanding of discount rates (the rate at which individuals trade current funds for future income). It makes sense to value money already in hand more than money potentially received in the future, but capitalism requires people to invest resources (usually money) now to make a profit later. Thus discount rates should reflect the profit one hopes to make in the future. Yet, householders have enthusiastically invested in housing with long-term returns being less than 10% (and the returns between 2006 and 2012 being negative), while largely ignoring investments in energy-efficiency improvements that give 50% or greater rates of return. Research shows surprisingly high discount rates for items ranging from refrigerators to retirement, which have spurred several high-profile efforts by corporations to entice employees to cash in early with reduced retirement benefits in exchange for waiving future benefits.[11] We suggest that learning about discount rates should be a fundamental part of K–12 education. Just imagine how much fun students would have if their math and science assignments included opportunities to teach their parents how to optimize the value of their home. Understanding discount rates has the potential to reduce the incidence of citizens being duped and swindled, at the same time that it promotes a more sustainable society.

The value of conservation can only be truly understood in the context of reaping future benefits associated with a functioning biosphere. If householders knew they could effectively increase their net worth by over 20% simply by choosing alternatives to commuting in their personal vehicles, and by making their homes more energy and water efficient, the evidence suggests that they would make more sustainable choices. Similarly, this awareness would spur the already-increasing value of homes that do not require automobiles or similar vehicles to reach shopping areas and places of employment, or do not need extensive remodeling to be energy efficient.

The second key to changing household policy/governance is competition. Electricity, gas, water, and other utility bills are typically treated as private information, but making them public would move household-level conservation into the realm of keeping up with the Joneses. Although some utility companies would make excuses about the need for privacy, there is no reason for them not to post neighborhood averages online immediately. Humans love to rank themselves, and rank appears to be more important in predicting happiness than overall wealth or well-being.[12] Modern householders' overwhelming desire to improve their relative social rank by purchasing prestige goods (large televisions, fancy cars, granite countertops) has been identified as a threat to sustainable long-term economic growth,[13] but this could be turned to more positive ends if data on the energy efficiency, walkability, and bikability of homes were available for householders to rank themselves with. Some preliminary research results suggest that this approach will work. The simple act of mailing utility customers a comparison of their electricity and gas consumption with that of "similar households" induced a 2.7% drop in energy use among study participants.[14] The approach was less costly (2.5 cents per kilowatt-hour) than most other options, and it could operate far more efficiently. Virtually all the expense of this approach could be eliminated if neighborhood data were made available online—just like other housing attributes neighbors enjoy perusing through popular websites such as Zillow (www.zillow.com)—and it would work better if the figures were for the homes of actual neighbors, because it would tap into the obsession modern society has for social ranking.

The Walk Score website (www.walkscore.com) taps into the practice of keeping up with the Joneses in terms of the walkability (and hopefully soon bikability) of homes. Users can enter their address and immediately find out how close most key amenities are to their houses, as well as find a rank for their home. Although the impact of this new, more sustainable form of ranking has not been evaluated, an unofficial experiment that one of us conducted (sitting down with colleagues and comparing the walkability of our homes) was compelling. Two homes were 1960s-style ranch homes near the center of Raleigh, one (ours) was a 1980s neo-Mediterranean house in a first-tier (close-in) suburb, and one in a second-tier suburb that bordered on McMansion status.[15] When we entered our addresses into the Walk Score website, the downtown colleagues gloated over their scores of 50–69 out of a possible 100. Because their homes were located near both shopping and work sites, their "somewhat walkable" scores surpassed our score of 49 (one point below "walkable") and dwarfed the score of the colleague with a home in a second-tier suburb, whose walkability score was a mere 5 out of 100. He had to settle for being labeled as "car dependent."

No, the colleague with a house in second-tier suburbia has not rushed out to buy a smaller home downtown, but that is not the point. As we mentioned at the beginning of this chapter, our goals are to make existing housing more livable, to appeal to rational self-interest, and to contribute to sustainable community development. The competitive spark lit by our little experiment has led to numerous conversations that go beyond our work group and have spread to our families and friends. Playing the walkability game, or even just talking about it, encourages people to start imagining options they had not previously considered when thinking about how to create more appealing neighborhoods.

Finally, if self-interest (in terms of take-home salary and social rank) fails to inspire change, fear also is a great motivator. Ulrich Beck noted that post-industrial societies were increasingly organized around an obsession with the future, especially in seeking to minimize risk and insecurity in that future.[16] This risk-averse society has driven people into suburbs built around the use of motor vehicles as they try to escape crime, ethnic diversity, and declining property values. Yet there are serious and rarely considered risks associated with living in and owning a huge suburban house connected to the necessities of survival only through one's car. James Kunstler has made a career out of his apocalyptic predictions about the abandonment of suburbia and the destruction of suburban communities when peak oil hits (the point when oil production reaches its maximum, and subsequently declines).[17]

Critics of these predictions note that, just like Paul Ehrlich's *The Population Bomb* and many other apocalyptic scenarios, the end has not come, and people still pay a premium for giant houses with few occupants far from urban centers by readjusting their spending patterns to allocate a higher percentage of their income to gasoline. Kunstler maintains a devoted following, however, because his predictions, and those of some other apocalyptic prophets, are logical (barring technological miracles). The green revolution and advances in biotechnology were the miracles that thus far have prevented the mass starvation predicted in Ehrlich's *Population Bomb*. What are the miracles to replace oil? Periodic disasters remind us that nuclear power is not the solution it once was believed to be, and alternative power sources (solar, wind, hydro, and geothermal) have limited applicability. Anyone who doubts the existence of a miracle cure for peak oil should think twice about energy efficiency and walkability scores before dismissing these as fads from the Far Left. Similarly, obesity has been linked to the first lifespan decline in 200 years for people living in the United States, and it is clearly tied to the walkability of home locations.[18] If householders in our modern society want to reduce their personal risk, they must realize that the perils of peak oil and obesity far outweigh the often-imagined risks associated with crime, crowding, and ethnic diversity in urban centers. Obesity and peak oil often are treated as political and social issues, but they pose very real menaces for individual householders, at least for those who are more likely to be seriously overweight and face financial challenges from inefficient homes and the expenses associated with a reliance on personal motor vehicles.

Community Solutions

Cities are relatively nimble governing bodies, with everything on the line when it comes to addressing the housing bomb. By 2010, most humans lived

in urban areas, including over 75% of the people in developed countries. This demographic transition makes humans an urban species, and cities are the key to any efforts promoting sustainable housing. Further, cities and counties constrain how development occurs at the neighborhood level, which is where almost all development decisions occur.

Cities face internal tension, since density and its ills must be balanced against the fact that density creates economic efficiencies that contribute to a city's competitiveness. Ultimately, density defines cities, increases their resiliency against the drastic swings in fossil-fuel prices associated with peak oil, makes them more sustainable, and facilitates a diversity of both uses and people.[19] The leaders of today's cities are aware that they risk being the next ring of suburbia to rot away as householders move farther afield or back into the revitalizing cores of major urban areas.[20] Some of the most innovative recent policies have emerged from their attempts to pioneer new approaches to traditional challenges associated with building a meaningful sense of community.

Places like Mesa (Arizona), Arlington (Texas), and Cary (North Carolina) are large and fast-growing, but they are not the metro centers in their regions, and thus are heavily automobile dependent. Unlike some suburban developments, however, ethnic diversity is relatively high, and places of employment are within their boundaries. These boomburgs use walkable outdoor minimalls to simulate traditional city centers, and it appears to work for the residents.[21] Similarly, boomburb mayors are rushing to develop light-rail systems, with nearly half having projects underway or with approved proposals, and over 80% expressing interest. Although installing a light-rail system seems to be about transportation (moving people from one place to another), the efforts of these community leaders have a more basic motivation. They want residents who have the option to live elsewhere to stay in their cities. A light-rail system is seen as a competitive advantage that established suburbs can offer over the sparkling new houses built in the next ring of development, located even farther from metro cores.[22] The pretend downtowns and the eagerness to install light rail highlight the fact that today's sprawl will be tomorrow's urban decay, unless such cities can make themselves more economically, socially, and environmentally sustainable.

In recent decades, the cities with the strongest economies have been populated with higher-than-average percentages of what Richard Florida calls the creative class: highly educated people who create new forms that can be reproduced, sold, and used (e.g., engineers, scientists, artists, entertainers, designers, writers, and analysts) or who engage in creative problem solving (e.g., those in high-tech sectors, the legal profession, health care, and business management).[23] These groups either create business or attract it, and they drive regional economic growth. They also move to places that embody innovation, diversity, and tolerance. The creative class isn't attracted to stadiums, sports teams, race tracks, or theme parks. As far as they are concerned, bigger is not necessarily better. They are drawn to high-quality experiences, including natural amenities (such as greenways), as well as places where they can readily participate in cultural activities and interact directly with one another, rather than having to deal with automotive travel. The creative class has begun clustering in areas with positive environmental and social attributes, and communities that develop green households and infrastructures will compete for these value-creating workers more successfully than those plagued by smog and toxins.

Principles for Promoting Sustainable Housing

Since the 1980s, several segments of the population that hope to address the ills of urban decay—including environmentalists, antisprawl groups, designers promoting neotraditional (human versus car-centered) development, and civil rights groups—have joined forces to promote sustainable cities and provided a litany of principles for achieving that goal. The first step cities should take toward encouraging sustainable housing is to create alternatives to Euclidean zoning and to traditional-subdivision regulations. Euclidean zoning is named for a 1926 Supreme Court case (*Village of Euclid v. Amber Realty Co.*) in which the court upheld the constitutionality of zoning. This allowed Euclid to prevent industrial development in nearby Cleveland from encroaching on the village, because it threatened the community. Since that time, Euclidean zoning and subdivision regulations have spread across virtually every city and county in the United States. They share three common attributes: (1) exclusive-use restrictions, (2) size and location restrictions, and (3) density restrictions.[24] Early in the history of zoning (the 1920s and before), use restrictions were stacked (one area allowing several uses): industrial zoning allowed commercial and residential development, commercial zoning allowed residential development, and so forth. Progressively, however, the model changed to clearly sort out all uses and prevent mixed use. Size restrictions (e.g., height) are often used to protect the rural character of an area, and location restrictions (e.g., required setbacks from the street or from other houses) address the character of the community and crowding issues. Density is controlled in several ways, including designating minimum lot sizes, minimum amounts of off-street parking, the number of units per acre, and the floor-to-area ratio (the ratio of lot size to the floor space of a building). This approach seems to encourage piecemeal development, and to omit any form of strategic planning. To be fair, the Standard Zoning Enabling Act of 1924 required local zoning to be based on a comprehensive plan. Unfortunately, comprehensive planning was never defined, and most courts have ruled that the mere existence of zoning is a plan in and of itself, so most states do not require planning as a basis for zoning. Euclidean zoning, as traditionally implemented, promotes sprawl and hamstrings efforts to arrange housing in a way that promotes the interests of either cities or residents.

The first place to turn to for alternatives to traditional-subdivision design is the urbanism principles pioneered by Jane Jacobs.[25] Jacobs convincingly argued that high density (of housing and activities), mixed uses (e.g., residential and commercial sites in the same street or even the same building), small-scale blocks and streets, and the retention of older and culturally valuable buildings all fostered economic and social diversity and sustainability. Her pioneering work has spawned the smart-growth, new-urbanist, and transit-oriented-development movements, as well as a host of additional principles for better urban planning. The *Charter of the New Urbanism* (www.cnu.org /charter/) advocates changing policy and development practices to reach several goals: (1) to create diversity in neighborhood uses and populations; (2) to design communities for pedestrians and public transit, in addition to motor vehicles; (3) to shape communities with clearly defined and accessible public spaces; and (4) to frame urban locales with designs and architecture that highlight local history, ecosystems, and climate. The smart-growth movement is quite similar to new urbanism, with perhaps less emphasis on areas that are already urban, and it has its own set of guiding principles:

1. Mix land uses
2. Design compact buildings
3. Offer a range of housing opportunities and choices
4. Create walkable neighborhoods
5. Foster distinctive, attractive communities with a strong sense of place
6. Preserve open space, farmland, natural beauty, and critical environmental areas
7. Strengthen and direct development toward existing communities

8. Provide a variety of transportation choices
9. Make development decisions predictable, fair, and cost effective
10. Encourage community and stakeholder collaboration in development decisions[26]

Advocates of transit-oriented development suggest organizing growth regionally (based on public transportation) and locally (based on the needs of pedestrians), an idea linked with Peter Calthorpe's pedestrian pocket: a subdivision within walking distance (often somewhat arbitrarily set at a quarter mile) of public transit, jobs, shopping, and parks.[27] A pedestrian pocket requires a certain amount of high-density housing and mixed uses (including commercial and retail establishments), to function correctly. Transit-oriented development also focuses on protecting sensitive wildlife habitat, riparian zones, and open space.

In 2009, the U.S. Green Building Council, the Congress for New Urbanism, and the Natural Resources Defense Council rolled out a new certification program for green neighborhoods: Leadership in Energy and Environmental Design (LEED) for Neighborhood Development. The program instigated a crucial step, both because it provided a concise summary of the rapidly expanding list of "smart" development practices, and because it expanded certification to the household sector, in addition to industry and government (which have far less room for sustainability improvements than households). The certification checklist (www.usgbc.org/DisplayPage .aspx?CMSPageID=148/) highlights three primary areas: sustainable locations and linkages, sustainable patterns and design, and sustainable houses. The location section focuses on reducing development in sensitive ecosystems and encouraging it on brownfields (abandoned industrial or commercial property). Among other things, this means avoiding development on wetlands, in critical wildlife habitat, or on historical sites. The LEED program provides lists of state agencies responsible for protecting these resources. Most of the states, in turn, have explicit maps of wildlife habitat and heritage sites, although these maps are woefully underused by developers. For example, the North Carolina Wildlife Resources Commission, through its Green Growth Toolbox program, provides free data on the locations of critical wildlife habitat, ecosystems, and heritage sites, as well as training on how to use this data. The linkages section of LEED for Neighborhood Development focuses on reducing people's dependence on personal motor vehicles through housing that is adjacent to jobs and shopping, connections to public transit, and the promotion of bicycling and walking. The sustainable pattern section focuses on walkability, connected communities, dense development, mixed incomes, mixed uses, reduced parking, and access to public spaces.

Most thought in the sustainable-houses sector has focused on new development, but many lessons can be drawn for the far larger sector of existing housing and neighborhoods. Communities can repair the damage caused by sprawl using the same principles advocated for the design of new communities and neighborhoods.[28] Some critics of renovating sprawling neighborhoods note that it is cheaper to design neighborhoods well in the first place, but if the cost of addressing urban decay is factored into the equation, rehabilitating preexisting areas becomes more appealing. If infill (using vacant sites in a built-up area) is planned in a manner that creates sufficient density to support public transit and introduces some retail and commercial development, existing periurban (single-function, usually residential) suburbs can be refashioned into walkable communities.

Although some smart-growth principles (e.g., replacing cul-de-sacs with connected streets, and promoting socioeconomic diversity) have proven nearly impossible to apply without policy interventions, its general principles for organizing households on the landscape are wildly successful. Having nearby parks and available sidewalks appear to increase community trust and social capital. Cul-de-sacs remain incredibly popular, however, despite the growing interest in traditional neighborhoods, and emerging research suggests that community trust (the basic building block of social capital) is actually higher in cul-de-sac streets than in traditional (connected) streets.[29] Somewhat ironically, the premium people pay for a cul-de-sac location reflects the high value they place on having a relatively vehicle-free environment, where social interactions outside of their houses aren't threatened by a frequent stream of traffic. This phenomenon clearly demonstrates why the market cannot be trusted to dictate development. Without policy intervention by cities, individual consumers are forced into a maddening game of creating car-based sprawl in their efforts to escape it. Similarly, market forces have not helped smart growth promote socioeconomic diversity. The success of carefully planned developments, and the relative lack of them, means that rapidly rising property values entice high-income buyers and motivate low-income sellers, creating communities populated by primarily white, well-off residents, despite the intentions of planners.[30]

Some householders will always demand wide-open spaces and will not find new-urbanist-style development appealing. For these individuals, conservation subdivisions provide a more environmentally friendly and sustainable alternative; they can even give residents more open space than is available in traditional large-lot subdivisions.[31] Conservation subdivisions retain undivided but otherwise buildable tracts of land as communal open space for their residents.[32] These subdivisions typically have the same number of housing units as traditional subdivisions, or even slightly more, but they cluster the units on 30%–50% of the land, leaving 50%–70% of the subdivision set aside as permanent open space. The open-space areas are delineated through site visits with developers and planners, and by an environmental inventory that highlights the most ecologically valuable land. This approach decreases landscape fragmentation; protects ecosystem services, including water quality, wildlife habitat, carbon sequestration, and aesthetic viewsheds; and can be planned regionally to create corridors for wildlife movement.[33] Better yet, in a similar housing market, houses in conservation subdivisions sell for higher prices than those in conventional subdivisions.[34] They can also be built for less money, because clustering development reduces infrastructure costs by 34%, compared with conventional subdivisions that require more grading, more storm-water infrastructure, and more roads.[35] Some conservation subdivisions are even being built with the intention of providing habitat for threatened species, like the South African oribi (*Ourebia ourebi*), a small antelope.[36]

Conservation subdivisions are surprisingly rare, given their promised environmental and economic benefits. This suggests that they face serious barriers beyond their novelty. Those barriers, in order of importance (as ranked by key stakeholders, including planning staff, developers, real estate agents, conservation groups, and private landowners), are lack of interest from developers, false perceptions that conservation subdivisions cost more to build, lack of support from elected officials, concerns about smaller lots, and issues with the long-term management of open space.[37] Expedited permits, density bonuses (allowances above the norm), reduced distance required for setbacks, and tax credits have proven successful in overcoming resistance from developers.[38] False assumptions about cost, elected officials' disinterest, and lot-size worries can all be responded to by educational efforts, including workshops and charrettes (intense, collaborative planning sessions).[39] Concerns about long-term open-space management can be addressed through a conservation easement (a transfer of development rights to a land trust or a governmental agency, guaranteeing that the open space would be conserved), and using stewardship funds and homeowner-association fees to cover open-space maintenance costs.

Communities also play a unique role related to energy: from power generation, through transmission and distribution, to use. Household-level power generation faces many challenges, including the fact that average householders will remain in a home for less than half the time it takes to recoup the costs of installing an energy-production system, and most energy systems have huge economies of scale (the larger the system, the less expensive it is to produce or acquire a product). Unlike individual homeowners, the average city should be around long enough to benefit from investments in sustainable forms of power generation, and cities can afford systems that are big enough to capitalize on economies of scale. Similarly, while individuals may find it difficult to borrow sufficient funds to pay for household-scale power-generation systems, generating systems represent a reasonable investment for cities. In Japan, communities investing in wind power not only have developed sustainable sources of power, but they have also promoted social cohesion and community commitment.[40]

Even moderate-sized cities can leverage resources to improve their energy profile. The Danish municipality of Frederikshavn bills itself as *Energibyen*, or Energy City. With a population of about 60,000, Frederikshavn aims to rely exclusively on renewable sources of energy by 2015, while also moving toward carbon neutrality, and it has instituted broad citizen participation to help reach that two-pronged goal (www.energycity.dk). In 2007, the municipality achieved a 20% reduction in its CO_2 emissions, accompanied by cost savings, simply by renovating municipal buildings. Frederikshavn's website informs visitors about current projects, including a pump that transfers heat from the sewage plant to warm approximately 400 households; a geothermal rooftop for the police station; and a solar system that produces energy for cooling and heating, as well as supplying hot water for its city hall. The challenges that have demanded the most coordination across different departments came from transportation. As one example, the neighboring town of Thisted reduced CO_2 emissions from its school buses by 10% by changing bus routes. Rather than having all buses arrive at the school at the same time, one bus runs a longer route, although this strategy required a change in the school's schedule.

Unlike municipal investments in roads and courthouses, community-based power generation can actually make a profit. Further, the cohesion created in the above examples is critical for communities hoping to secure a vibrant future in a context where citizens, who provide economic growth and a tax base, are highly mobile.

Overcoming Barriers to Sustainable Housing

The first suite of tools needed to promote sustainable housing addresses strategic changes in housing density. Cities need ways to decrease density in areas with historical, agricultural, and ecological value and increase it elsewhere. This essential ability is stymied in the United States, where property is considered a naturalized (inviolable and presocietal) right, because constraining density changes the value of one's property. Although specific policy tools have been designed to avoid clashes based on property rights, perhaps the most important strategy is planning growth before it becomes a high-stakes game. When property values are low, and property owners aren't contemplating multimillion-dollar payouts from cashing in on their property investments, community-wide development plans can be made and codified in a way that addresses community interests without the lurking prospect of major sums of money soon to change hands, and with less likelihood of prompting angry complaints about property-rights violations. Once developers set their sights on acquiring specific properties, compromise, good will, and collaboration become difficult to achieve, even with the best policy and planning tools.

Project-specific reviews of development are the low-hanging fruits (the ones easiest to reach) serving as an alternative to traditional zoning. They allow developers to bring a planned unit development (PUD)—generally proposing mixed uses, a higher-than-normal density accompanied by protected open space and/or greenways, and other innovative features—before the local planning board and attempt to receive either a special-use permit for the development or a zoning amendment for it. The PUD process promotes community collaboration and creates flexibility, but it requires time and money. Accordingly, it is used relatively rarely, typically only for large development projects, and generally only when communities promise incentives to developers, such as bonuses allowing variances in local standards regulating the height or density of buildings.[41]

A more comprehensive solution is to adopt new zoning regulations that permit more sustainable neighborhoods on an as-of-right basis. Cities throughout the United States, and several states, have instituted new zoning codes.[42] These codes rely on mixed-use zoning, minimum-density require-ments, limits on the maximum permissible amount of parking, minimum open-space requirements, and rules to protect agricultural land. They are collectively known as traditional neighborhood development (TND) codes. The codes, which allow innovative and sustainable housing on an as-of-right basis for development, are more effective than other incentives that fail to accomplish the same goals.[43] The *Conservation Subdivision Handbook* (www .ces.ncsu.edu/forestry/pdf/ag/ag742.pdf) provides model zoning-code lan-guage (adapted from conservation-subdivision pioneer Randall Arendt) for conservation subdivisions.

Addressing problems associated with low density, isolated uses, and de facto segregation in cities will require more than simply instituting new zoning regulations based on principles from the new urbanism, smart growth, and transit-oriented-development movements. The legacy of Euclidean zoning means that these types of planned growth need help, and cities have sev-eral tools available to encourage developments that follow TND codes rather than traditional zoning, including quotas, moritoriums, concurrency, growth boundaries, and inclusionary zoning. Quotas and moritoriums (delays or waiting periods) are difficult to justify, unless cities can demonstrate they are required to avert disaster (e.g., to ensure timely hurricane evacuations, or to prevent disease). The other tools are typically supported by the courts, at least when they are backed by planning and clearly not intended to completely stop development.[44] Concurrency ordinances require public infrastructures (e.g., transportation and utilities) to be available to mitigate impacts from new de-velopment.[45] Growth boundaries have proven to be a controversial tool for controlling the density of development, however. Urban-growth boundaries can take many forms, but all focus on defining city boundaries, promoting greater density within the city, and advocating low density outside the city. They are discussed in more detail in chapter 7, because (with few exceptions) they necessitate regional and state-level coordination.

The final tool, inclusionary zoning, requires builders to construct a mini-mum percentage of affordable housing, and it is typically limited to large proj-ects.[46] Inclusionary zoning reflects a response to the dominant practice of exclu-sionary zoning. Exclusionary zoning is used to maintain large minimum-lot sizes and minimum-floor areas, while discouraging mobile homes and multi-family housing in an area. Inclusionary-zoning programs avoid lawsuits by providing developers with alternatives, either by allowing affordable housing to be constructed off site, or by imposing fees in lieu of on-site affordable housing. Inclusionary zoning nonetheless faces some challenges. It can drive up housing prices by reducing their supply; builders can circumvent the intentions of this form of zoning by bundling mandatory payments for amenities (e.g., golf courses) with home purchases; and relatively low in-lieu fees allow builders to avoid constructing mixed-income housing.[47] It does appear, however, that mandatory inclusionary-zoning programs are more effective than other affordable-housing programs, and in recent decades inclusionary zoning has not hampered the housing supply.[48] Nonetheless, because exclusionary zoning is used by communities to keep out lower-income families, a regional approach, versus one at a city level, may be needed.[49] Cities do have an impetus to utilize inclusionary zoning as a way to prevent urban decay in existing neighborhoods, or when a lack of affordable housing creates labor shortages.

The promotion of public transportation, walking, and bicycling are some of the most powerful tools municipalities have for making housing sustainable, and they can be among the cheapest. Transportation systems that replace personal motor vehicles with rail systems (light rail, subways, and commuter trains), buses, bicycles, and walking save lives, save money, and save the environment.[50] Research has shown that the exhaust from motor vehicles kills nearly three times as many people in the United Kingdom as vehicular accidents.[51] The affordability of integrated transportation systems is highlighted by the fact that Bogotá, Colombia, and Curitiba, Brazil, have led the way in converting transportation by automobiles and similar vehicles to forms dominated by pedestrians, bicycles, and buses.[52] The key to the success of using buses in these two areas was a rapid-transit system that sped up bus travel relative to that in personal motor vehicles, and it has been emulated throughout the world, in developing and developed countries alike. These changes have been linked to declining crime rates and enhanced economic vitality, and they have helped make the streets of Bogotá safer than those of Washington, D.C. Bicycle infrastructure, in particular, is inexpensive. Between 2007 and 2009, Seville, Spain, installed an 87-mile bicycle network for $43 million.[53] While not trivial, the cost of the Seville system was about half the price of repaving just three miles of interstate highway in Los Angeles ($75 million).[54] The percentage of trips made by bicycle in Seville increased from 0.1% in 2000 to 7% in 2009, and the number of personal motor vehicles in the city's historic district fell by over 50%.

In Portland, Oregon, the urban-growth boundary (and the resulting increase in density) has been less instrumental in promoting sustainable transportation between homes and other destinations than changes in that city's transportation infrastructure.[55] Compact neighborhoods did not measurably reduce driving, but mixed-use neighborhoods, the provision of public-transit services, and decreased accessibility to freeway interchanges limited the amount of single-person driving. Interestingly, research in Oregon suggested that providing public-transit services and creating mixed uses for lands near homes was more effective in reducing dependence on personal motor vehicles than providing transit services and mixed uses around workplaces and commercial areas.[56]

Relatively small investments in bicycling infrastructure can have huge economic returns for communities, both in terms of increasing housing values and attracting consumer spending.[57] In North Carolina's Outer Banks, a $6.7 million investment in bicycle infrastructure now generates $60 million annually in bicycle tourism. In this region, adjacency (roughly three-tenths of a mile) to bike paths created an 11% increase in home values. Similarly, adjacency to public transportation (e.g., light rail) raised home values in other areas.[58] Preliminary research from a wide range of communities suggests that immediate adjacency (less than a quarter mile) increased commercial property values by 16%, and residential property values by 4%. Farther away, the dominant effect is on residential homes, with a 2.3% increase in property values for every 250 meters (820 feet) closer the property is to a rail station. Although virtually all concerted efforts to promote bicycling—improved infrastructure (e.g., dedicated bicycle lanes), pro-bicycle programs (e.g., safe-route-to-school programs, and cash incentives), and restrictions on the use of personal motor vehicles (e.g., tolls, or temporary road closures to motorized traffic)—have proven successful, no single approach has been ideal.[59] Similarly, cities as diverse as Bogotá, Barcelona, Berlin, and Boulder have doubled or tripled their share of bicycle trips over the last three decades, along with reducing injuries and fatalities. Unfortunately, most U.S. cities have a horrible track record when it comes to protecting bicyclists, with mortality rates about double those in other developed nations, and injury rates between 8 and 30 times higher.[60] The good news for "car countries" like the United States and Australia, which are heavily dependent on the use of large personal motor vehicles, is that with a 1% share of trips by bicycle (versus a 30% share for bicycling nations), and a heavy bias toward males (75% of bicycling is done by males in car countries, versus an even gender split for bicycling nations), there is plenty of room to capitalize on the efficiencies and sustainability associated with alternatives to cars and similar vehicles, and ample opportunities to bring those alternatives to children, women, and the elderly.

Finally, integrated transportation serving households will keep the elderly and the poor from being imprisoned in their homes. The generation currently reaching its golden years has grown accustomed to unprecedented mobility, activity, and freedom, but it faces an inability to find satisfactory alternatives to the driving associated with most current housing.[61] The newly elderly face being cut off from the amenities purportedly making their retirement better than that of previous generations. A survey of seniors in Southern California found they fear losing their ability to drive more than losing a spouse or facing a child's poor health.[62] This result becomes less surprising when one realizes that trends toward having fewer children and not living with those children mean that when many of the elderly lose their ability to drive, they will be alone in neighborhoods that have no mass transit, and no safe way to walk or bicycle.[63] The erratic but ever-increasing cost of fossil fuels compounds this problem for poor people and those with fixed incomes (often the elderly). The drastic decline in miles driven since the global recession in 2007, irrespective of gas prices, has occurred without comparable increases in miles travelled by bicycle or by mass transit.[64] The conclusion is simple: reliance on a housing infrastructure based on personal motor vehicles is robbing people of their freedom, and it starts with the elderly and the poor.

Most traditional-neighborhood developments with mixed uses (housing plus retail and commercial establishments) and reasonable density have created high consumer demand, but mixing types of people has proven to be in low demand. In Orenco Station, a new-urbanist development in the greater Portland, Oregon, area, two-thirds of the residents were happy with the current level of diversity (95% white) and opposed to the development of affordable housing nearby.[65] Although Portland was able to generate greater neighborhood satisfaction in its high-density and mixed-use neighborhoods relative to Charlotte, North Carolina, residents in both regions were happiest in single-family, detached housing.[66] Based on these preferences, addressing problems associated with low-density, isolated uses and de facto segregation in cities will require more than leveling the playing field with new zoning.

Any efforts to accomplish more than giving everyone the same opportunities, however, face two primary barriers: property-rights complaints and homevoting. Any form of urban planning has the potential to impact property rights by changing property values. Indeed, the urban planning that privileges motor vehicles (e.g., investing in roads) has increased the values of property far from cites and decreased property values in core urban areas for half a century. People who own more-remote properties seem to feel entitled to the continuation of planning practices that increase their property values.

The good news is that planning that contributes to just small losses in property values is not considered to be a taking by U.S. courts, which means that municipalities are not required to reimburse landowners for lost value in those cases. When planning drastically reduces a property's value, cities can either pay for the taking or utilize tools similar to a transfer of development rights (TDR).[67] Using TDR, a municipality can protect environmental, historical, or agricultural lands by preventing development. Landowners denied the ability to develop their own property, however, still have as many development rights as other landowners and can sell them to developers planning to build in receiving areas (places where development is desirable). If receiving areas are not available, the development rights (credits) can be sold to a bank, which holds them and sells them at a later date.

Certain behaviors by homeowners, dubbed the homevoter phenomenon, create a major barrier to any innovations in planning where households can be located on the landscape. According to the homevoter hypothesis, householders' homes are their largest assets, so their votes in local politics are based on the immediate economic interests related to their homes.[68] This has been interpreted to mean that homevoters' political choices are to stop all change, because any change is a gamble on the value of their most substantial asset. The safest strategy for protecting their investment is pulling up the gangplank after they are aboard and preventing any changes. If homeowners cannot succeed by voting at the ballot box, they vote with their feet and leave, creating urban decay. Homevoting has helped prevent the introduction of environmentally dangerous development (e.g., polluting industries), but homevoters have also opposed nearly all other changes, ranging from apartments and similar high-density developments to additional high-priced housing and greenways.[69]

Cities cannot foster innovation, diversity, and tolerance (which constitute the recipe for building the creative class and for economic prosperity) without promoting change, and change is not possible without the support of homevoters. So how can cities calm panicked homevoters? Insuring a householder's largest asset is a reasonable start in promoting more rational decisions. In many ways, houses are more like investments in the stock market than purchases of traditional commodities. When prices fall for automobiles, demand goes up, but when prices drop for homes, people panic and communities can die. A rush to the suburbs is like a rush on the banks, and insurance provides a partial solution. Home-value insurance was used to prevent panic selling in Oak Park, Illinois (near Chicago), when racial integration was imminent in the 1970s.[70] Since then, no claims on that insurance have been filed, strong housing markets have prevailed, racially diverse communities have developed, and mixed incomes (and a solid tax base) have persisted.[71] Similar results have been achieved in other cases, including other locales around Chicago, military communities, and Syracuse, New York.[72] These programs suggest that this form of insurance serves to prevent irrational panic, and if that panic can be avoided home values can increase in the face of various changes: racial integration, the inclusion of some high-density housing, the creation or expansion of mass transit, and other modifications that make urban economies more efficient and urban living more desirable, relative to the next ring of suburbia. Home-value insurance can take many forms and must be tailored to local contexts, but the methods used, possible legal issues, and context-specific concerns have been clearly laid out for communities that are interested in innovation but afraid of homevoters.[73]

Although facilitating sustainable transportation at the local level (e.g., walking, biking, and city-wide transit) is a fundamental responsibility of cities, it can be a daunting task. As with individual householders who lack the resources to research all the options available to them, most municipalities lack the resources to attain a clear understanding of the markets, costs, and benefits associated with developing infrastructures for walking, cycling, and public transit. Fortunately, these critical pieces of information were recently compiled by the National Transportation Research Board.[74] Handy online tools (e.g., www.ecu.edu/picostcalc/) are also available for municipalities and businesses who want to evaluate the markets, costs, and benefits associated with providing bicycle and pedestrian infrastructure. Users simply need to enter their locale and then provide information on the type of infrastructure they are considering, their location, and some basic census data (e.g., median income, population density). The federal government and state governments also have incentives to support municipal efforts to create sustainable transportation (see chapter 7).

Some critics of smart growth suggest that suburban sprawl is actually cheaper to support than more dense, walkable, well-thought-out developments. These naysayers, however, fail to consider other differences between the two scenarios. Their findings are generally reported outside peer-reviewed literature, where they are less subject to serious scrutiny. The typical approach is to compare the costs of governance in older, dense urban cities to those of governance in new suburbs and merrily point out that dense urban areas have higher costs. This approach is flawed for several reasons. First, portions of many existing metro areas have been abandoned by fleeing middle- and upper-class residents, leaving an unusually high density of vulnerable residents who need expensive social services from cities. Thus the determinants of city budgets are being driven by factors beyond infrastructure and sprawl. Similarly, most large urban areas utilize a system of government based on paid employees, while newer sprawling suburban areas, even those with populations exceeding 300,000, often have volunteer mayors and rely on homeowners' associations (to which members pay fees) to fulfill the roles traditionally taken on by city government.[75] These in-kind payments (volunteered time and services), as well as homeowners' association dues, become resources used for local governance, and they should be included in cost estimates. If differences in governance style and social services are controlled when modeling costs for cities and suburbs, planning for future development, rather than letting it proceed unchecked, undeniably saves money and resources, and improves human well-being.[76]

The most legitimate criticism of well-planned housing development is that it is so popular that demand quickly makes prices too expensive for low-income householders and excludes minorities.[77] Yet this problem, too, has been overcome in several instances. About 25% of the new-urbanist developments in Minneapolis–St. Paul consist of affordable housing, and affordable housing has been constructed in conservation subdivisions, such as Blue Sky Acres, built by Habitat for Humanity (a nonprofit organization) in Hickory, North Carolina.[78] This subdivision is the most popular Habitat for Humanity community in the city, despite the 0.07-hectare (0.17-acre) lots. The fact that people strongly prefer a more sustainable pattern for household development should encourage all of us to experiment with options for expanding its availability, rather than discourage us from implementing it.

Conclusion

Although state and national policy places limits on what cities can do, the evidence shows that the flexibility of cities enables them to experiment with new forms more readily than states, nations, or international bodies. This finding should revitalize our efforts to envision new possibilities for households. In fact, especially on issues such as climate change, cities have emerged as the leading force for spurring action.[79] People have been moving to cities ever since they figured out how to accumulate sufficient excess production of food and goods to sustain urbanization. Although the process varies across time and space, the overall trend toward increased urbanization shows no signs of weakening. People move to cities in search of opportunities, social connections, and access to innovation. Initiatives that harness these motives offer the most immediately practical ways to defuse the housing bomb.

CHAPTER SEVEN

Large-Scale Strategies
for Defusing
the Housing Bomb

State governments, other regional governments, nations, and even the international community have a critical role to play when it comes to defusing the housing bomb. Many issues associated with where housing goes on the landscape (e.g., health, or social justice) traditionally fall outside the purview of local governments. Communities with good intentions to coordinate smart growth are often stymied in their efforts by regional development patterns beyond their control, including white flight and a race to the bottom where lax communities focus primarily on the short-term profits of attracting development and tax dollars.[1] Also, the aggregated results of smart growth planned at the community level do not necessarily combine to protect watersheds, airsheds, habitat connectivity, viable transportation networks, or a stable climate. The regional patterns resulting from community-planned smart growth may not be very smart when they are considered at a regional level. Celebration Florida provides a great example of this phenomenon. This master-planned community was designed using new-urbanism principles, but it was located in one of the last wildlife corridors connecting north and south Florida. When the few remaining open spaces between Orlando and Tampa are gone, terrestrial wildlife will find south Florida sealed off from the rest of the continental United States.[2]

Households have become a huge resource sink. Even before they are built, homes and subdivisions start sucking in energy and spewing out emissions, some of which have been linked directly to anthropogenic (change related to or resulting from the influence of humans on nature) aspects of climate change, primarily from CO_2, and others that are connected with more immediate health effects (e.g., sulfur and nitrogen oxides). Changes in lifestyles, in home construction, and in municipal planning can contribute to energy conservation, but to go further, we need to switch to low-carbon energy sources. Although much of our energy use occurs at the level of individual buildings and structures, the options available to their owners and to communities are strongly influenced by state and regional policies. Further, state and regional energy policies are linked to national and international contexts. Emerging energy technologies in the United States are funneled into subnational state energy systems that are, in turn, embedded in national and global energy systems.

State and Regional Solutions

States have enormous economic incentives to promote both more sustainable distributions of housing, and more sustainable forms of transportation to link housing with workplaces and amenities. The most obvious incentives are reducing health-care costs and increasing economic competitiveness. States share the costs associated with Medicaid and the Children's Health Insurance Program with the federal government. According to the National Association of State Budget Officers (www.nasbo.org), those costs had climbed to $130 billion a year in 2010. The $130 billion constitutes 13% of the total for state budgets, less than that allocated for education, but double any other spending category. Further, while education costs for states (but not families) have remained stable for over 20 years, states' health-care expenditures have been climbing, due to both increasing costs per person and increasing enrollment.

Schools provide a great example of why states need to promote the complete-streets movement (www.completestreets.org). The basic idea behind complete streets is designing and operating streets to promote their safe use by pedestrians, bicyclists, motorists, and transit users of all ages. Between 1970 and 2000, the percentage of children walking or biking to school at least one day a week fell from nearly half (48%) to less than 15%.[3] The most important barrier to children walking to school was distance; the next was traffic. Nonetheless, nearly half of the children who still live within a mile of their schools are active travelers (as opposed to passive riders), walking or bicycling to school. The southern United States, however, lags about 20% behind other regions of the country in terms of active travel to school. Promoting active leisure (e.g., sports) is no substitute for designing landscapes where active travel to and from school, work, and other daily destinations is possible. The health benefits of bicycling to work have been shown to reduce mortality risks by 40%, even after accounting for leisure-time physical activity and other demographic variables.[4]

Allowing municipalities to scramble for the tax revenues associated with sprawling subdivisions and strip malls can be a dangerous gamble for states that rely on tourism dollars or on natural amenities that contribute to a state's quality of life. The states that have aggressively developed growth-management programs have diverse political leanings, but they are united by their abundant natural resources, tourism based on the state's natural amenities, and expansive population growth[5] The list of states is growing, and early innovators include Wyoming, Georgia, Florida, Hawaii, Maine, and Oregon. The housing bomb is no respecter of party politics; it can strike in blue, red, and purple states. If unplanned development fouls the waters of southern Florida, tourists will fly to the Bahamas; if subdivisions gobble up critical winter range for elk in Wyoming, hunters will drive across the state line to Idaho or Montana.

States also influence which energy options make sense for households. Each state has a unique combination of natural resources, political constituencies, and an existing energy system that shapes how new energy technologies are integrated into it. State and regional institutions determine the capacities of an energy system, including whether or not various energy technologies are utilized. This, in turn, influences the options that make sense for householders and developers to consider. If a local utility company is required to buy excess energy back from individual householders, then solar roof panels on their homes will pay for themselves much quicker. Analyses of energy technologies used in the United States have found that states arrived at their current mix of technologies via quite different cultural, economic, and political paths.[6] This helps explain why and how technologies were—or were not—made available to individual householders.

Principles for Promoting Sustainable Housing

Despite the diversity among states, Jay Wickersham notes that in their efforts to plan ways to deal with housing development and its associated transportation networks, those states that have enacted growth-management statutes have reached a strong consensus on eight principles:

1. Protect natural resources
2. Improve, or protect, water quality
3. Preserve farmlands and forests
4. Preserve historic landmarks
5. Protect open space

6. Promote economic development
7. Develop transportation systems that include alternatives to personal motor vehicles
8. Create, or protect, affordable housing[7]

Overcoming Barriers to Sustainable Housing

The states that have chosen to directly address sustainable housing have developed five primary models for influencing local development: (1) state or regional land-use classifications, (2) state or regional regulations for large development projects, (3) state or regional regulations regarding critical resources, (4) investments in state infrastructure in locations where growth is desired, and (5) state or regional requirements that local land use be consistent with state or regional goals.[8] Hawaii led the approach involving state-level land-use classifications. Their broad urban, rural, agricultural, and conservation classifications create a framework under which communities use more detailed zoning to plan development. Most states engaged in regional planning require state or regional approval for large development projects that surpass thresholds deemed to create regional impacts. The Florida Land and Water Management Act of 1972 gave the state power to designate areas of critical concern, and to issue special planning and development guidelines for them. Since that time, the City of Apalachicola, Key West (as well as all the other islands in the Florida Keys), the Green Swamp, and the Big Cypress Swamp have been designated as areas of critical concern. To avoid promoting sprawl, Maryland has successfully steered state investments in roads, schools, colleges, and offices into existing developments.[9]

Planning consistency, the last of these five models for state involvement in local development, is the most comprehensive, useful, and controversial. The basic approach, pioneered in Oregon, requires local plans to conform to statewide goals, and local development to conform to the local plan. Consistency between local regulations and state goals can be difficult to determine, however, unless the state develops specific mechanisms for achieving those goals, such as urban-growth boundaries or minimum housing-affordability standards. Urban-growth boundaries clearly define growth areas, where density and development are encouraged, as well as conservation areas, where growth is discouraged and critical resources (e.g., farmland, endangered-species habitat, or historical landmarks) are protected.

The first U.S. urban-growth boundary was established in Lexington, Kentucky, in 1958, but the Portland, Oregon, urban-growth boundary (created in 1980) has been the most famous, as well as the most notorious.[10] This boundary has been highly controversial, and early research suggested that it failed to slow sprawl. The rate of transit use (eleventh overall) had grown faster than in other metro areas, but not much faster than the rate of private-vehicle use (twelfth overall). This failure was due in part to a regional flaw in the system, where sprawl simply flowed across the Columbia River to Clark County in Washington, where Oregon regulations did not apply. After subsequent regulations in Clark County closed the release valve, the Portland urban-growth boundary has worked more effectively.

Efforts to promote sustainable housing by state and regional agencies face the same primary areas of resistance as those occurring at the community level: property-rights concerns and homevoting. In a backlash against perceived property-rights violations associated with state and regional planning, Oregon voters passed Measure 37 in 2004. Measure 37 required state or local compensation for any reduction in property value associated with land-use regulations (versus the traditional approach of only requiring compensation for an almost complete loss of value), and it allowed property owners to ignore most planning regulations if they did not receive their cash within two years of filing a complaint. The proponents chose a 92-year-old woman, Dorothy English, as their property-rights icon, and trumpeted the dramatic story of planning regulations denying her the ability to subdivide her land and give it to her children. After Measure 37 was exposed as gutting any regulatory ability to plan for future development, Measure 49 was passed by a wide margin in 2007, and it has come close to removing the restrictions of Measure 37. The short-lived success of Measure 37 inspired similar efforts in Arizona, California, Idaho, Montana, Nevada, and Washington, which were all defeated or invalidated by the courts. A highly modified version that focused primarily on eminent domain was eventually enacted in Nevada.

Legal support for urban-growth boundaries has solidified, and interstate collaboration (such as that between Oregon and Washington in the Portland area and California and Nevada around Lake Tahoe) can help prevent the leapfrog phenomenon, where householders create super sprawl by bypassing areas protected by the urban-growth boundary and driving farther out, until they come to a locale where they can afford a house. The key is to protect land far enough away from the urban core that leapfrogging no longer makes sense.

Regional coalitions of local municipalities, planning firms, and conservation-minded non-governmental organizations (NGOs), such as land trusts or The Nature Conservancy, have developed another alternative for stopping leapfrog development.[11] Homeowners interested in promoting housing values can foster exclusivity and much higher home values by artificially creating a built-out scenario (a condition where additional development is no longer economically viable) through mandatory large lot sizes (e.g., 20 acres) and the protection of significant areas of open space (either through land trusts or conservation subdivisions). Thus sizeable areas with no remaining pockets of land large enough to support economically viable developments could be created quickly. Communities, including those with residents from diverse socioeconomic backgrounds, often adopt the artificial build-out approach when they see neighbors using it. When requirements for large lot sizes are combined with aggressive efforts by land trusts to protect open space, sub-division development rapidly loses its economic viability: the number of homes per development crashes, and developers must redirect their build-ing efforts back toward urban cores, where higher-density construction is allowed and small-parcel development is still profitable. As one example, this phenomenon appears to be happening in the New Jersey Highlands.[12] Some may criticize this approach, because it is unaffordable for low- and middle-income families, but—as this book demonstrates—owning a home in subur-bia may mean economic disaster for such families. The real danger with this planning model is an inadequate acquisition of easements that protect open space, with the resulting super-sprawl engulfing the entire landscape.

Current markets create a barrier to environmentally and socially sustain-able housing, because markets, since the late 1800s, have been rigged in ways that promote sprawl.[13] Land trusts and conservation easements, however, found a niche in that system in the early 1980s. The Tax Reform Act of 1976 provided the first federal legislation formally recognizing conservation ease-ments as tax-deductible gifts.[14] Although conservation easements were used in the 1880s to protect parkways in the Boston area, and in the 1930s to pre-serve scenic vistas along the Blue Ridge and Natchez Trace Parkways in the southeastern United States, the 1976 tax reforms helped set off a flood of tax-deductible easements. The Uniform Conservation Easement Act was final-ized in 1981 and used in the laws of 23 states, while 26 states wrote their own version of the statute. Between 1988 and 2005, the number of land trusts more than doubled (from 743 to 1,537), and the amount of land that was protected skyrocketed from 290,000 to over 5 million acres. The National Conservation Easement Database (http://nced.conservationregistry.org), released in 2011, indicated that the quantity of conservation-easement land has continued to grow exponentially; it was at 20 million acres by 2010. Conservation easements have been made even more desirable by federal estate-tax deductions, and by several states that allow income- and property-tax reductions.

Conservation easements are appealing for several reasons. They keep land on the tax rolls, protect only critical values of that land, are cheaper than outright purchase, allow landowners to maintain some property rights, and provide lucrative tax breaks for landowners. Two notable criticisms of conservation easements are that they are growing too fast for land trusts to keep up with their management, and that the incentives are far larger for wealthy landowners than for land-rich, but cash-poor, landowners.[15] In 2004, an average high-income landowner would have paid $133,237 less in taxes by placing a piece of land valued at $500,000 in a conservation easement, but a middle-income and a low-income landowner would only have paid $36,450 and $9,450 less, respectively.[16] The regressive nature of tax benefits (benefits that decrease for less wealthy landowners) associated with conservation easements is all the more painful for low-income landowners who want to protect their land, because they have the highest opportunity costs (the value of foregone possibilities) associated with giving up the right to develop their property further or engage in other lucrative land uses. Creating progressive incentives (benefits that increase for less wealthy landowners) for conservation easements would both broaden the market and be more socially just.

Promoting affordable housing and socioeconomic integration are also controversial measures, because they conflict with homevoter tendencies in places where communities and neighborhoods exclude poor people and minorities; both of these groups are perceived as threats to the local tax base and property values. Integration, however, is in states' interests, because integrated communities and schools can help people lift themselves up by their bootstraps, while segregated development (with pockets of abject poverty) create huge expenses for the states' social services. Massachusetts passed an "antisnob zoning act" (Massachusetts Comprehensive Permit Act, chapter 40B) in 1969 to address this issue. The law allows builders of affordable housing to override local zoning bylaws and other requirements in any municipality containing less than 10% affordable housing. Developers also receive density bonuses (allowances above the norm) if 20%–25% of their new units have long-term affordability requirements. Several other states (including Connecticut, New Jersey, and Oregon) have followed these efforts, with slight modifications. For example, the New Jersey Fair Housing Act of 1985 required municipalities to submit plans describing how they would provide a fair share of affordable housing, and to demonstrate that they were actively following those plans.

When it comes to overcoming unsustainable energy use, a smart grid offers the most significant technological transformation of energy use since the introduction of electricity into individual households.[17] A smart grid is a digitally enabled electrical grid that gathers, distributes, and acts on information about the behavior of all participants (including suppliers and consumers). It represents a complex set of technologies with the potential to reduce the costs and enhance the efficiency, reliability, and sustainability of electrical production, storage, transmission, distribution, and use. Smart-grid systems are critical to developing a sustainable energy system. Although significant variation is apparent in visions of what these systems are and how they are developing, smart grids are characterized by abilities to

- Heal themselves
- Motivate consumers to actively participate in grid operations
- Resist attack
- Provide higher-quality power to prevent outages
- Accommodate multiple generation and storage options
- Enable electricity markets to flourish
- Permit a higher penetration of intermittent power-generation sources

Industrial customers already use partial versions of these metering systems. These customers know the cost of electrical power during different times of the day, and they are able to alter their power use (e.g., cycle furnaces on or off) based on the price of energy. Households consume far more power and have much greater room for efficiency gains than other sectors, so extending the empowering aspects of smart-grid technology to householders represents a crucial step in making an energy system more efficient, more reliable, and better able to accommodate renewable energy.

Householders can be integrated into a smart grid in two primary ways. First, their homes can be connected to the grid through new meters that have greater capacities, including the ability to show energy prices in real time. This step is essential, because it gives householders the option of using electrical power during those times of the day when it is cheapest. Most householders, however, will not watch the meter, with prices shifting up and down every few seconds, before deciding when to push the start button on the dishwasher. Therefore the second crucial step is promoting smart appliances. Smart appliances monitor electricity prices and then use power when it is the cheapest. National support is vital in encouraging the adoption of smart appliances, and it can operate in the same way as Energy Star labeling and rebates, which have successfully promoted the adoption of energy-efficient appliances.

Extending a smart grid to households will save householders money and make energy markets more efficient, but it can also benefit the environment by helping renewable-power systems, based on wind and solar energy, achieve sufficient reliability for broad-scale use. Because wind and sunlight are intermittent electrical-generation sources, they typically require extensive supplemental power from coal or natural gas plants to achieve sufficient reliability.[18] Turning these backup plants on and off creates inefficiency, but this problem can be reduced if the smart grid is used to dampen swings in power demand through a more efficient timing of appliance use. If battery power (e.g., in vehicles) is integrated into smart grids, a smart grid can even respond to drops in renewable energy production by pulling power from households (and paying residents for that power).[19] Tying houses to smart grids can make renewable energy viable in places where it would not otherwise be so, and enhance the reliability of renewable-energy power supplies.

For power distributors, a smart grid means using automation to improve their control over and operation of the distribution system. Electronic sensors and remote-control switches can limit the extent of power outages by rapidly identifying and isolating a problem from the rest of the system. These sensors and switches have the ability to selectively shed loads (specific sites on the grid that use electricity), so critical services (such as hospitals) could continue to receive power, even in an emergency that reduces the supply of electricity below the demand.

The transmission of electricity from power plants to users also can be made more efficient. Although some instrumentation already is used to monitor power flows along transmission corridors, measurements at key locations could significantly cut the amount of energy that is wasted during transmission. Other devices can respond to short-term peaks and valleys in power demand by changing the electrical properties of the power lines, sending power in one direction, rather than another. Transmission systems are currently controlled centrally, using a combination of human operators and computer systems. A smart-transmission system would replace this centralized control with distributed control systems, which would enable a more efficient use of information to fix problems before power consumers are even aware that they have occurred.

If smart grids will do all these things, why don't we have them everywhere? Although many of the technologies needed to roll out smart grids are available, significant economic and social barriers remain.[20] In Europe and the United States, these impediments include regulatory environments that don't reward utilities for operational efficiency, the limited ability of utilities to transform their business model and take advantage of smart-grid technologies, and the challenges associated with computer security. Consumer advocates have expressed concerns over personal privacy, the establishment of a fair basis for the availability of electricity, private-sector abuses (e.g., insider trading and other forms of information leverage), and additional governmental control over the use of electricity.

Substantial smart-grid rollout depends on the organizations responsible for moving power across space. In the United States, these organizations are usually called Regional Transmission Organizations (RTOs) or Independent System Operators (ISOs). A few states (California, New York, and Texas) operate their own systems, with minimal interaction across state boundaries. If developers or municipalities want to initiate smart-grid projects, they will be working with the RTO/ISO where they are located. Vermont was among the earliest states to invest in smart grids, via a state initiative in 2004. Vermont's smart-grid projects operate in cooperation with ISO–New England. When the American Recovery and Reinvestment Act of 2009 provided funding, smart-grid projects developed throughout the nation (www.sgiclearinghouse.org /ProjectMap/). North Carolina's Department of Commerce highlights smart grids on its website (www.energync.net/government-nfps/energy-assurance -program/smart-grid-technologies/), where viewers can learn what this concept means by clicking on links such as "Smart Grid 101" or "The Smart Home." The demonstration project centered in Raleigh, North Carolina, works in cooperation with PJM, a regional transmission organization that is responsible for moving electricity across 13 states and the District of Columbia. Pecan Street, Inc. has a smart-grid demonstration project sited in Austin, Texas, which functions in cooperation with the Electric Reliability Council of Texas (www.edf.org/energy/building-smarter-grid-austin-texas/). The microgrid project in Borrego Springs, California, operates in cooperation with the California ISO (www.smartgridlibrary.com/tag/borrego-springs/). Other projects similarly depend on cooperation with the RTO/ISO located in their general area. These regional organizations ensure that power gets from one place to another, and they support the demonstration projects' efforts to implement a vision that seems appropriate to their particular locales.

National Solutions

The United States, perhaps more than any other nation, should be motivated to promote sustainable housing. This country helped trigger the 2007 global recession, due to the decades-long bubble in its housing prices. By pioneering sprawl, the United States has suffered some of the worst per-house ecological damage among developed nations. The distribution of housing in the United States also makes this nation among the most vulnerable in the world to peak oil, and to any other kind of energy shortage.[21] Global trends in energy use, however, give the United States ample hope of reversing this trend. First, the amount of energy used for private transport declines asymptotically as urban density increases.[22] For example, a relatively small jump in density between Atlanta (about 11 people per hectare [2.74 acres]) and London (about 60 people per hectare) reduces personal-transport energy use by 80,000 megajoules per year, while a further jump in density from, say, London to Hong Kong (about 270 people per hectare) only produces a modest decrease of 10,000 megajoules. U.S. cities can achieve drastic reductions in their energy demands more easily than any others in the world. U.S. Energy Information Administration data (www.eia.gov) notes that the lack of density in the United States is a major reason this nation ranks below most other developed countries in terms of the energy efficiency of its economic output. In 2006, the United States used about 9,000 BTUs per dollar of its gross domestic product, relative to 7,000 BTUs in Europe. The efficiency drag on the economy is minimized when energy is inexpensive in all of its forms, but when a critical form (e.g., oil) skyrockets in price, countries with the lowest efficiencies suffer the most. The United States is also more vulnerable to peak oil than developing countries, because what little economic output the latter have requires very little energy, and their households are growing denser, while the United States is beset by sprawl.

The federal government should provide incentives for and subsidize municipal efforts to create complete streets—facilitating movement between homes and other destinations by pedestrians and bicycles, in addition to personal vehicles—because obesity is set to become an enormous drag on human health and the economy. Countries that promote sustainable transportation

Table 7.1 Obesity rates in Organisation for Economic Co-operation and Development nations in 2005

Nation	Percentage overweight (BMI >25)		Percentage obese (BMI >30)	
	Men	Women	Men	Women
USA	69	58	32	34
England	62	52	21	22
Australia	60	42	18	16
Canada	59	40	17	14
Spain	53	34	12	10
Austria	50	33	11	10
France	43	28	9	9
Italy	47	27	8	7
Korea	35	26	3	3

between homes, workplaces, and other destinations will have a competitive edge economically. The effects of the obesity epidemic (table 7.1) are no less astounding than the potential effects of peak oil. Among the 34 Organisation for Economic Co-operation and Development nations, the United States is the fattest, and only trails Kuwait, some of the Pacific Islands, and Saudi Arabia in terms of the percentage of its obese and overweight citizens.[23] Recent research suggests that if obesity trends from 1970 to 2004 continue into the future, over 86% of Americans will be overweight or obese by 2030, and all Americans would be overweight by 2048.[24] The total health-care cost in the United States attributed to obesity is doubling every decade, and it is projected to account for 20% (nearly $1 trillion per year) of all U.S. health-care costs by 2030. Obviously something will need to change before the last healthy American succumbs to weight problems, but the trend is astounding. Urban planners and transportation experts can connect homes to other destinations in ways that promote active lifestyles, enable citizens to live healthier lives, and help countries like the United States cut into the projected $2 trillion annual costs of obesity and peak oil.

Principles for Promoting Sustainable Housing

Although municipalities and states have generated a remarkable amount of agreement on how to promote sustainable housing, recent trends suggest that the U.S. federal government couldn't agree on principles for tying a pair of shoes. Not surprisingly, there are not mutually agreed-upon principles for federal action to defuse the housing bomb, so we proffer a somewhat right-wing policy principle: get out of the way and let states and cities promote sustainable housing. Of course this principle is more difficult to follow than it sounds, because the federal government currently plays central policy and financial roles in many aspects of housing, ranging from how homes are built and financed to how they are connected to the rest of the world through transportation, electricity, and communications infrastructures. In the short term, states, cities, and individual householders must rely on progressive federal policies to successfully promote their own economic, social, and environmental well-being. In effect, "getting out of the way" equates to developing federal policies that reflect the consensus-based decisions of states, cities, and householders. Given the recent upswing in political polarization, it may seem like a stretch to suggest there are consensus principles for the federal government to follow, but a second look at the aforementioned city and state principles may counter that reaction:

- Preserve open space, farmlands, natural beauty, and historic landmarks
- Improve, or protect, environmental quality
- Promote economic development
- Develop transportation systems that include safe alternatives to personal motor vehicles
- Create a range of quality housing opportunities and choices
- Strengthen and direct development toward existing communities
- Make development decisions predictable, fair, and cost effective
- Encourage community and stakeholder collaboration in development decisions

These principles embody widely shared goals. Virtually everyone wants to save farms and beautiful landscapes, and preserve their local history. Everyone wants clean air and water. Everyone wants growing and vibrant economies. Everyone wants their children, themselves, and their grandparents to be able to safely walk down the street without being crushed by an automobile. Everyone wants choices in quality housing. The difficult parts are deciding how to achieve these goals and determining who will pay for the required efforts.

Overcoming Barriers to Sustainable Housing

The federal government has three basic tools for influencing citizens—the carrot (economic incentives), the stick (regulation), and the sermon (information)—and they can all be leveraged to defuse the housing bomb.[25]

Using the carrot and the sermon are actually related, because capitalism only works efficiently when certain assumptions are met, one of those being that consumers should have complete information about the products they consume, their alternatives, and the costs for both. The federal government plays a vital role in informing the public about the sustainability of housing. Current federal programs (e.g., Energy Star and WaterSense) focus on new homes and appliances. The U.S. Green Building Council's LEED efforts to promote energy-saving green buildings demonstrate that NGOs can play an important role in informing the public, but these programs focus on new construction and commercial and industrial buildings, where far less opportunity for energy efficiency occurs than in the existing-home sector. European nations have pioneered an approach for addressing this need with their Energy Performance Assessment for Existing Dwellings (EPA-ED) program.[26] The EPA-ED program already has clearly articulated methods and software available for uniform energy audits. The program gives prospective home buyers savings estimates, specific advice on energy-performance improvements, and an energy-performance certificate.

Similar approaches in the transportation sector would improve the sustainability of housing. The effective commuting speed of motor vehicles is typically lower than that of bicycles when the time spent working to pay for automobile-related costs (not just how long a person is on the road going to and from the workplace) is factored in. Currently, the label (window-sticker certificate) accompanying a new vehicle shows that vehicle's estimated mileage per gallon and compares its fuel costs or savings with that of an average vehicle. There is no customization, and the label fits the same make and model of every vehicle in the nation. With little or no cost, that label could be customized to a specific buyer. Carfax-type programs help buyers view the history of individual vehicles instantly, at very little cost. That model could be easily flipped to allow buyers to get a more accurate understanding of how a particular vehicle would help or hinder their efforts to address their transportation needs. A revised certificate could factor in insurance costs, local taxes, maintenance costs, parking costs, interest costs, the locale where the buyer lives and works, and even impacts on personal income.

The automotive industry would fight such measures, primarily because they know householders underestimate the overall costs of driving by 60% or more, and they overestimate their commuting speed relative to mass transit and bicycling, but such changes would allow the market to be more efficient by removing information barriers and providing equal opportunities for those industries associated with mass transit and bicycling. Further, the approach would be in the best interests of consumers and the environment. One obvious complaint would be related to how much personal information is required to create an individualized vehicle label, but given the fact that financing for typical vehicle sales requires all of the proposed information (and more), this complaint rings hollow.

Providing better information is only the first way the market's power can be harnessed to address the housing bomb. The second strategy requires removing governmental subsidies that reward irresponsible behaviors. There is little motivation for individuals to make responsible choices when they can count on property-tax relief, protection from bankruptcy proceedings, and forgiveness of excessive debt on the homes they have purchased. And lenders are only behaving logically when they advertise the availability of these subsidies.

The federal government can use the carrot approach outside the marketplace in instances where crucial amenities needed by the public are not adequately addressed by markets. National defense, education, energy, and transportation are sectors where most moderate Americans agree that federal support is mandatory. Energy and transportation are the most immediately relevant in terms of the housing bomb. The federal government's support for smart-grid demonstration projects illustrates what can be accomplished when a state provides appropriate encouragement for new initiatives. For example, the federally supported Pecan Street project in Austin, Texas, has grown into an extensive consortium that integrates multiple private partners, ranging from SunEdison (solar energy production) to Whirlpool (home appliances). The consortium brings together researchers, marketers, and power companies, along with residents of the community. And the neighborhood is being built on an abandoned airport.

The federal government also can promote sustainable housing by encouraging sustainable transportation to and from that housing. Sustainable housing requires both making private motor vehicles more efficient, and making alternatives to them more viable. Less governmental help is needed in terms of making automobiles and similar personal vehicles more efficient, because rapidly rising fuel prices are already starting to propel efficiency gains. Since infrastructure and policy has stacked the deck against mass transit, pedestrians, and bicycles since the early 1900s, greater governmental assistance is needed to level the playing field for alternatives to personal vehicles. The complete-streets movement has capitalized on both the need for viable alternatives to driving and the lack of fairness in how transportation infrastructures are designed.

Data from the National Cooperative Highway Research Program (www .fhwa.dot.gov/research/partnership/nchrp/) notes that support from several federal sources for pedestrian and bicycle-friendly infrastructures has grown rapidly since the early 1990s. The Clean Air Act Amendments of 1990 helped create the Congestion Mitigation and Air Quality Improvement Program, which has progressively provided more support for alternative transportation infrastructures (increasing from $3 million in 1992 to $97 million in 2011) in its efforts to improve air quality. Other efforts include the Safety Set-Aside (1992–2003) and Highway Safety Improvement (2004–2012) Programs. The latter started funding pedestrian and bicycling safety efforts with their first $2 million investment in 2004, which crept up to $5 million in 2011. The broader Surface Transportation Program set-aside for Transportation Enhancement Activities began funding pedestrian and cycling infrastructures with $6 million in 1992, which had increased to $90 million in 2011. The Safe Routes to School and Nonmotorized Transportation Pilot Programs started in 2006, with $17 million invested in infrastructure designed to promote walking and cycling among children in grades K–12 by ensuring that they had safe routes to school. Funding for these programs increased to $130 million by 2011, but they were placed on the chopping block in 2012 by partisan squabbling in Congress. The budget for the Recreational Trails Program grew from $3 million dollars in 1999 to $31 million in 2011. After all of the miscellaneous federal projects supporting pedestrian and cycling infrastructures are accounted for, federal funding increased from $23 million in 1992 to $790 million in 2011. This upswing may seem dramatic, but it has a long way to go, since the total national effort to make pedestrian and bicycle travel safe and viable costs about the same amount as repaving 30 miles of urban interstate highway.[27]

In the long term, the federal government needs dedicated funding for promoting pedestrian and bicycle-friendly infrastructures; in the short term, the available funding comes primarily from fuel taxes. Some critics may argue that governmental funding generated through taxes on drivers should only be allocated to projects focused on improving infrastructures for gasoline-powered vehicles, or that the federal focus should be on interstate modes of transport (rarely foot traffic or bicycling). For multiple reasons, those arguments are incredibly short sighted and narrow in scope. First, in border areas between states, walking and bicycle riding can be of vital importance to interstate travel. Second, promoting a greater mode share (the number of individuals using a particular form of transportation) for pedestrian and bicycle travel helps meet clearly articulated federal responsibilities for protecting clean air and water. Third, investing in pedestrian and bicycle transport meets the needs of people paying taxes on motor-vehicle fuels better than investing in motorized transport systems.[28]

Roads and highways between states would have lower maintenance costs and less need for expansion if a significant mode share is converted from motor vehicles to pedestrian and bicycle travel. Because the costs of creating a 10%–20% mode share are miniscule relative to the expansion and maintenance of car-based infrastructures, investing in pedestrian and bicycle infrastructures is a better use of tax revenues from motor-vehicle fuels, even if improving the driving experience for motorists (e.g., better roads, or less traffic congestion) is the sole measure of success. For example, the Canadian province of Québec developed one of the world's premier bicycle systems—2,702 miles of bicycle paths and road routes—for $95 million dollars in 2000, a modest sum in comparison with standard highway costs.[29]

In the United States, the Nonmotorized Transportation Pilot Program was perhaps even more successful.[30] The program allocated $25 million annually to four areas (Columbia, Missouri; Marin County, California; Minneapolis, Minnesota; and Sheboygan County, Wisconsin) to create infrastructures supporting pedestrian and bicycle travel. Between 2007 and 2010, the programs diverted 16 million miles of what would otherwise have been motorized travel to walking or bicycling. These communities experienced a 49% increase in the number of cyclists and a 22% increase in the number of pedestrians. Bicycling mode share grew by 36%, pedestrian mode share rose by 14%, and driving mode share decreased by 3%. Because the number of fatal crashes involving pedestrians and bicycles dropped during that period, the decrease in the amount of motorized trips reduced the economic cost of mortality by $6.9 million. Finally, critical expertise was built among planners in these communities, and they now have long-term plans to continue their support for complete streets.

These results clearly demonstrate that minimal prompting at the federal level can rapidly create huge gains for complete streets. The United States must capitalize on this potential, because the distribution of housing relative to schools, retail establishments, and jobs makes our nation the most vulnerable to peak oil, and we have the highest reliance, by far, on motorized vehicles of any nation in the world.[31] European countries not only greatly exceed the United States in terms of walking and bicycling, but they are also increasing the mode share of walking and bicycling trips faster than the United States.[32] This is no small feat, since increasing the share of walking from 40% to 41% is much more difficult than increasing that share from 1% to 2%. Further, active travel is distributed more or less equally by gender and age groups in Germany, but women, children, and the elderly actively travel far less than men in the United States. In 2009, Germans walked more than twice as often, cycled more than six times as frequently, and walked at least 30 minutes a day more than three times as often (29%) as did Americans (8.8%).[33] The latter statistic may be the most troubling in terms of health-care costs, because exercising for 30 minutes a day has been linked to reduced risks for many chronic health conditions. Investing in sustainable infrastructures connecting houses to other destinations would be worthwhile just by being the most efficient way to improve road conditions for motor-vehicle users. The aforementioned benefits of tackling the projected trillion-dollar annual costs of obesity and the trillion-dollar annual costs of peak oil, as well as protecting the mobility of children and the elderly are merely icing on the cake, but they make a nice, thick layer.

Although relying on the stick (enforcing governmental regulations) is often frowned upon, it has a stellar track record in terms of promoting sustainable housing.[34] The United States has largely taken a piecemeal approach (regulating consumer products, appliances, lighting, furnaces, and fixtures) versus addressing overall energy use in homes. The regulations stem from the 1975 Energy Policy and Conservation Act (EPCA), the 1987 National Appliance Energy Conservation Act, and a series of amendments to those acts (the most recent being in 2007). Today's EPCA provisions require standards to be the highest that can be economically justified with current technology.[35] Washing machines and dishwashers were regulated in the late 1980s; refrigerators and freezers, air conditioners, water heaters, furnaces, boilers, cooking products, and pool heaters were regulated between 1990 and 1992. Faucets, showerheads, and toilets were regulated in 1994. Lighting was regulated intermittently (as new products were developed) from 1990 to 2012.

In 2010, all of these regulations saved 3 quads (short-scale quadrillions of BTUs) of energy, and they had saved 26 quads since 1987. A quad is a unit of energy equal to 1.055×10^{18} joules, about the same amount of energy contained by 8 trillion gallons of gasoline. Additionally, these regulations saved every U.S. consumer at least $175 in 2010. As has been the case throughout this book, residential savings from these regulations were more important than commercial and industrial savings. In fact, regulations created four times as many savings in the residential sector as in the industrial and commercial sectors combined.[36] Water-conserving standards saved 1.5 trillion gallons in 2010 (9% of all public-water withdrawals), and they had saved 11.7 trillion gallons since 1987. In 2010, the same regulations accounted for 167 million tons of reduced CO_2 emissions. These successful regulations are paralleled in the transportation sector by Corporate Average Fuel Economy (CAFE) standards. CAFE standards were first created in 1975, and they pushed fuel economy about 12% higher than it would otherwise have been by 2002.[37] Such standards are crucial for the very reason why critics decry them. Some detractors suggest that these standards make vehicles more dangerous by requiring that they be lighter and smaller, but driving is safer in Europe than the United States, precisely because most vehicles embody these two attributes. Vehicular lightness, particularly in the minds of Americans, is often associated with cheapness (and thus poor construction), but that is not necessarily the case. Nor does crash-worthiness have to be compromised. Formula One racing has proven that lightweight vehicles (weighing roughly 1,300 pounds) can survive accidents at very high speeds, protecting their drivers far better than the vehicles available to consumers. The tendency for fearful U.S. drivers is to use larger and larger vehicles in efforts to ensure that they will run over whomever they collide with, and CAFE standards address at least a portion of this issue.

The dramatic success of the regulatory approach suggests that continuing along the path of periodic updates in the standards for appliances and furnishings in homes will play an important role in the sustainability of houses, but the United States could benefit from adding other elements to those rules, focusing on the efficiency of entire homes.[38] These regulations could require that minimum standards be met, both when homes are constructed and when they change hands through purchase or rental. For instance, in Germany homeowners are allowed to choose their preferred method for achieving energy efficiency (e.g., more insulation versus better heating technology).

The U.S. federal government has also effectively used the stick approach to make household locations more sustainable, with the Endangered Species Act of 1973 preventing home development on critical endangered-species habitat. Similarly, President George H. W. Bush's adoption of a "no net loss" policy for wetlands under the Clean Water Act has made it much more difficult to pave over wetlands. These regulations could be augmented by following the lead of states that direct infrastructure funding in ways that promote sustainability, such as geographically distributing transportation funds based on existing housing density, rather than on connecting new (sprawling) communities to existing areas of density, and giving priority funding to projects that provide safe, non-motorized transportation options for moving between homes, schools, places of employment, shopping areas, and recreational sites.

International Solutions

The international community has a critical role to play in defusing the housing bomb. Housing challenges in developed nations (such as sprawl, dependence on automobile-based infrastructures, obesity, and vulnerability to peak oil) are surpassed by challenges in developing nations. The speed of urbanization has only been exceeded by the speed of urban-slum growth in developing nations.[39] The United Nations estimated that in 2001, 1 billion humans were living in slums around major urban centers, and that number is on pace to reach 2 billion by 2030. In other words, in 2030 one quarter of humanity will live in substandard housing, with little or no tenure security. These homes pose disease risks, lack sanitation, lack clean water, lack adequate energy, and are incredibly vulnerable to natural disasters.

According to the United Nation's 2011 *Millennium Development Goals Report*, between 2000 and 2010, nearly 200 million slum dwellers experienced improved conditions in their water supply or sanitation, or obtained more durable housing, and the percentage of urban householders in developing nations who lived in slums dropped from 39% to 33%.[40] While these advances are important, they failed to keep pace with the explosive growth of urban centers in developing countries, and the overall number of people living in slums increased by about 100 million. The projected movement of over 1.8 billion people into urban areas in the developing world (from 3.2 billion in 2003 to 5 billion in 2030) makes addressing housing in urban slums a critical international need.

The persistence of slums in urban areas throughout the globe does offer unprecedented opportunities to defuse the housing bomb. By their nature, new urban slums are extremely dense, facilitate pedestrian use (because residents do not have automotive vehicles), and provide the only viable entrance into urban areas for poor rural people. While the abject poverty of many urban slums is disturbing to many, urban poverty is a highly visible displacement of less-visible rural poverty. The improvements in slum housing for 200 million residents during the last decade reflects the incredible success that is possible with concerted efforts to provide land security, paved access, drainage, sanitation, water, and financial credit for low-income families. The United Nations' *Slum Upgrading Facility Handbook* describes a detailed process whereby urban areas can help householders living in slums improve their communities.[41] Making these communities more livable could become the biggest contribution to sustainable housing in human history by shifting the human-settlement patterns for billions from sprawling communities to dense and walkable urban areas.

Principles for Promoting Sustainable Housing

Unfortunately, the *Slum Upgrading Facility Handbook* also reveals its acquiescence to the prevailing neoliberal dogma that views human organization as no more than a collection of actors that each function independently from the whole.[42] If humanity is composed of rational agents, acting purely out of immediate self-interest, it is reasonable for the handbook to describe low-wage householders in slums as "the gold-dust of the urban economy" and slums as "the gold-mine of the city economy."[43] While we cannot disagree with the claim that workers living in slums provide their cities and nations with a competitive advantage in terms of minimal labor costs, their economic value should not be reduced to merely being a means of production for capitalist profits. And their communities are more than sites to be mined for the economic advantage of the wealthy. This mindset poses a serious threat to the possibility of diffusing the housing bomb by improving housing in urban slums.

Although living conditions in slums can be rapidly improved, their residents are innovative, independent, and valuable human beings who are often treated as nothing more than resources that can be exploited for short-term economic gain.[44] In the last two decades, millions of people per year have been evicted from urban slums, despite the fact that such evictions are a clear violation of international law.[45] These evictions are typically justified by the need to build infrastructures (e.g., roads), by master-planning for growing urban areas, or as preparation for international events (e.g., the Olympics), but they invariably hurt slum dwellers and benefit local-growth coalitions. Although there is nothing inherently untoward in enriching members of local-growth coalitions, there is something deeply wrong with destroying slum dwellers' social networks, their livelihood strategies, and their property, as well as with the assault, rape, and murder that frequently accompany these forced evictions.

Urban renewal, especially when accompanied by political repression, has long been an instrument where the housing sector provides stability for crisis-prone economies. Baron Haussmann's destruction and rebuilding of Paris in the 1850s, the suburbanization of the United States in the 1950s, and the urbanization of China occurring over the past twenty years have all helped solve surplus-capital and unemployment problems.[46] In those cases where we have evaluated the results, the housing bomb has not been defused. Instead, in cities throughout the world (e.g., Seoul, Mumbai, Beijing, London, and New York), slum dwellers have been removed, their property has been expropriated, and the newly available spaces have become sites for the accumulation of more profits by the already wealthy.

David Harvey notes that, when combined with authoritarian control, neoliberal logic paints these campaigns as necessary, and even positive, in their economic and environmental outcomes.[47] In Seoul in the 1990s, developers hired thugs to drive homeowners out of neighborhoods that had been built in the 1950s; the developers then knocked down the houses and built the elegant skyscrapers that characterize the skyline today.[48] In Mumbai, financial institutions are arranging for the removal of millions of people who have been living as squatters in slums that have now become desirable real estate. In Beijing, approximately three million people were evicted from their homes as part of the urban renewal in preparation for the 2008 Olympics.[49] In other cities, such as London and New York, the removal of the urban poor has been less violent, but just as effective. And, although these slum clearances enable cities to report a drop in the percentage of householders living in slums, the people that inhabited them have not disappeared. Instead, they have set up camp along roadsides, reestablished new slums just beyond the old boundaries, or simply become homeless.[50]

Overcoming Barriers to Sustainable Housing

Fortunately, there are alternative approaches that could guide more humane, and ultimately more sustainable, ways to transform urban environments. The Healthy Cities movement suggests an approach that may enable us to loosen the hold of neoliberal logic. It depicts cities as complex systems and advocates short-term experiments, grounded in local conditions, supported by the state and (when available) by private investment, but controlled by democratic forums of local residents.[51] Proponents note that, although humanity has made progress toward the United Nations' Millennium Development goals, millions of urban residents still lack access to basic sanitation, and they experience dangerously high levels of air pollution (both indoors and outdoors) on a daily basis. Although, in the aggregate, the generally accepted belief that national income per person correlates with urban health may hold true, the correlation does not hold up when applied to the urban poor.[52] This finding may be explained by the increasing differences between the living conditions of the very wealthy and the very poor that have come to characterize society in the twenty-first century.[53]

The *Lancet* Commission's report on urban health suggests that the best results (in terms of improving the everyday lives of people) will be achieved by attending to these differences, and adopting small-scale experiments focusing on options that can be independently sustained by the very poor.[54] They offer the example of 328 toilet blocks that were constructed throughout poor neighborhoods and now provide sanitary access for an estimated 400,000 slum dwellers. This project may not seem to be connected to housing, but it is directly relevant; it illustrates the importance of working within the existing system, rather than violently evicting householders over and over again. In chapters 2 and 6, we described the proliferation of household appliances that characterize suburbia, and we also noted the savings in energy and other resources that can come when householders share such appliances. Just as a householder in a Stockholm apartment may prefer to share the purchase and upkeep costs of laundry facilities with other members of the housing cooperative, householders in a Mumbai neighborhood may prefer to share the purchase and upkeep costs of toilet facilities. Multiple small-scale projects that are frequently evaluated, and that can be managed by local residents, may offer effective (if painstaking) approaches to defusing the housing bomb internationally.

Transportation and mobility are also critical issues in the world's urban slums. Slum dwellers may not emit as many greenhouse gases as their wealthy neighbors, but this difference is not necessarily their preference. People who are poor walk (or ride bicycles), because other forms of transportation are either unavailable or too expensive. And this phenomenon illustrates the global nature of today's society. The three of us live in different regions of the United States, and we all know people who bike (or walk) to work because they lack sufficient funds for other forms of travel. Although (in most cases) they cannot afford to live close enough to their places of employment for biking to be convenient, they also do not have enough money to purchase a motor vehicle, and public transportation is not available. These workers are sending most of their earnings to individuals in their families who live in slum conditions in relatively poor countries and have no other means of support. Thus money earned in Colorado, Michigan, New York, or Texas will often support family members living in the slums of Mexico City, Mumbai, Rio, and San Salvador. Even at this scale, the global nature of housing dynamics should be clear. This is not to say that solutions will come easily, but rather to emphasize that they matter to us all.

The importance of small-scale experiments that are tailored to local conditions cannot be overemphasized. Even something as seemingly minor as a community-subsidized bicycle-repair shop can play an important role for people who rely on bicycles for transport to and from their places of employment. Traffic-calming strategies also encourage people to continue using bicycles, even when other transportation alternatives become available, because such strategies contribute to the perception of safety. This further adds to individual human health by encouraging the continued incorporation of physical activity into one's daily routine, and by avoiding the greater quantity of emissions that accompany increased motor-vehicle traffic.[55]

Finally, the advantages of smart-grid energy systems are not limited to wealthy cities, states, or nations. It is only reasonable to expect that people who currently live in inadequate housing will seek to improve their standard of living, and along with that improvement come increased demands for electricity. The World Health Organization (WHO) estimates that low- and middle-income urban residents incur a larger health burden from indoor air pollution than from outdoor air pollution.[56] Since heating and cooking with less-clean fuels causes the majority of this burden, WHO advocates switching to cleaner fuel, which often means electrification.[57] Yet increased electrification means greater greenhouse-gas emissions, and, often, other outdoor pollutants (depending on the process used to produce the electricity). The development of smart grids for electricity in these instances offers the same advantages already demonstrated for wealthy nations and cities. The Global Smart Grid Federation (whose members include Australia, European Union countries, Great Britain, India, Ireland, Israel, Japan, Korea, and the United States) investigates challenges and opportunities for the utilization of smart grids in the highly diverse environments represented by its members. Their reports summarize ongoing initiatives and explore new possibilities for meeting human energy-supply needs without the levels of environmental degradation that have accompanied previous expansions of the energy sector.[58]

Conclusion

The complexity that globalization has injected into housing makes the neo-liberal tendency to reduce human society to nothing more than a mass of independent actors especially dangerous. Technological developments that enable us to substitute one resource for another (e.g., energy can be provided by coal, petroleum, or even the human "gold dust" living in slums) discourage us from paying attention to and dealing with the risks of our decisions. When humans are reduced to nothing more than a resource that can transform raw materials into commodities to be sold in the marketplace, their essential connectedness with us as fellow humans is concealed, and it becomes easy to ignore injustice.[59] A quick glance at the armed conflicts erupting around the world, however, should remind us that a failure to attend to risks does not make them go away. One ironic result of globalization is that unanticipated threats to sustainable human habitation can easily emerge from choices made in a completely different part of the planet. This means that addictive housing behaviors in any nation may set off housing bombs in apparently unconnected countries. Along with this danger, however, comes the potential that even incremental attempts to defuse the housing bomb can reverberate throughout multiple sites. The suggestions in this chapter illustrate how policy instruments drawn from many levels of governance can support incremental efforts to develop more sustainable housing.

Conclusion

Given that prices for homes registered their largest drop in history on the December 30, 2008, Case-Shiller home price index report, our warning about a housing bomb should not be surprising. In the spirit of Paul Ehrlich's bold predictions in *The Population Bomb*, we predict that the social, the economic, and particularly the environmental impacts of the 2008 price implosion will pale in comparison with future disasters if communities, states, and nations fail to defuse the housing bomb before the demographic transition in average global household sizes reaches 2.5 people per household (chapter 1). Escalations in environmental degradation, an economic decline, and obesity-related health problems will plague any community, state, or nation that fails to address the housing bomb. There will be exceptions in the short term, such as fossil-fuel boom towns, but ultimately even these communities will end up as tired, worn out, dirty relics unless they address the housing bomb before the stores of gas or oil run dry.

The Apollo 8 astronauts helped launch environmentalism by taking the world's most famous photograph in 1968: an iconic image of Earth rising above the moon's horizon. For many people, this image encapsulated the fragile and finite nature of humanity's home in the universe. Most of the diverse groups that made up the 1970s-era environmental movement, however, focused on the fragile and finite nature of Earth, rather than on the idea of home.[1] The specters of warfare, economic crashes, and the mass starvation that would ensue when the growing human population overwhelmed the limited resources of our planet defined the environmentalist perspective toward population-control policies, debates over immigration, and even the preservation of endangered species. The overpopulation argument has currently lost momentum, however, since the direst predictions associated with it have not materialized. Although population numbers may underlie social, economic, and environmental challenges today, few people think addressing population alone will solve the problems associated with consumption.

Even though population growth has been flattening out, human relationships with nature are becoming progressively more destructive.[2] As the human condition improves, population growth tends to decline. The opposite relationship holds true for households. As the human condition improves, households get physically larger, are placed in or adjacent to natural areas more frequently, require more transportation energy, and provide shelter for fewer people per housing unit.

Fundamental changes in culture, economics, education, and technology have combined to drastically alter household dynamics (chapter 1). In an ironic twist on the attempt to moderate the rate of population growth, aging has led to a drastic rise in the number of households per capita. Other cultural phenomena, including increasing divorce rates and the decreasing incidence of multigenerational households, also contribute to household proliferation. Household proliferation constitutes an important threat to biodiversity conservation. Each home typically brings with it an entourage of unexpected effects, including the three most important drivers of the current mass extinction of native plant and wildlife species: habitat loss, the introduction of exotic (nonnative) plant and animal species, and landscape fragmentation.[3] Of course, a large number of people who neither understand nor care about the importance of biodiversity to human survival have no qualms in paving over wildlife habitat. But even people who support biodiversity conservation and place a high value on wildlife threaten key habitat with their housing choices (chapter 3). This group is more likely than others to locate their houses in places that endanger the very species they value.[4]

The increased wealth of many demographic groups has led to larger footprints for homes, magnifying the effects of more households that contain fewer persons. In the United States, the average physical size of a housing unit more than doubled during the twentieth century. Industrial and postindustrial economic development and globalization have combined to free household-location decisions from their traditional constraints. In the United States, for example, air conditioning facilitated massive displacement from the colder Rustbelt states to the warmer Sunbelt ones. Widely available devices, and low-cost and fast communication systems, are now helping to create a communication-based economy, with a growing workforce capable of living anywhere. These facts, combined with a likely preference for ecologically sensitive areas among wealthy, highly educated, and environmentally oriented people, presents unprecedented conservation challenges.[5] The skyrocketing per capita consumption of resources represents a ticking time bomb, and, as this book demonstrates, households sit at the nexus of consumption: of natural resources, of energy, of water, and of land itself. Accordingly, households represent the key to addressing economic stagnation and decline, social disparity, biodiversity loss, and climate change.

Despite Garrett Hardin's claim that focusing on the human population as the problem was taboo,[6] the emphasis on overpopulation has a long history, dating back at least to Malthus's *An Essay on the Principle of Population*, which was first published in 1798. Malthus's arguments about overpopulation played a key role in the rhetoric justifying the atrocities of Social Darwinism and eugenics, arguments for forced sterilization, opposition to immigration by 1970s-era environmentalists, and China's one-child policy.[7] Focusing on population hasn't been taboo for the past generation. In many ways the subject of controlling houses is more taboo than controlling population. The passion for property rights and the tendency to equate home ownership with virtue and democracy (chapters 2 and 3) has stymied efforts to promote sustainable household dynamics for the past several decades. Similarly, loss of the freedoms associated with home ownership has weakened the ability of working-class Americans to demand change. Most Americans have been convinced that they need to own a home, but in the aftermath of the mortgage crisis and the Great Recession, they are only a few house payments away from losing their most significant asset, and thus cannot afford to negotiate with employers about their jobs or salaries, or take financial risks to create their own employment (chapter 2).

To defuse the housing bomb, we must overcome the current taboo against community planning, and recognize that our systems of taxation, our permitting regulations, and our loan guarantees have skewed markets toward unplanned sprawl (chapter 2). Our nation's skyrocketing economic vulnerability associated with peak oil (the point when oil production reaches its maximum, and subsequently declines); obscenely high obesity rates, linked in part to automobile-based lifestyles; the associated loss of mobility among those who cannot drive; the massive loss of productive farmland; and the declining livability of American cities (chapter 6) leave no other choice.

The refreshing difference between a population bomb and a housing bomb is that with the latter, solutions are abundant, are everywhere, are relatively easy to implement, and create immediate change (chapters 3–7). To make things even better, the taboo against challenging free-for-all housing development has set the stage for what could be a cascade of sustainability improvements with regard to homes. The United States now has the least-sustainable stock of housing in the developed world in terms of energy, water, and land use, and there are abundant examples of how these choices can be changed quickly and with minimal resources. Teton Valley in Idaho (chapter 3) and Wolong Nature Reserve in China (chapter 4) highlight how communities can turn the tide on sprawl and housing proliferation, and point out permanent scars left by waiting too long before defusing the housing bomb. These two case studies also emphasize crucial principles for communities in similar situations: address planning before property values soar, recognize that property rights are not absolute, and use conservation-planning policies that are appropriate for the local culture and local contexts.

Individual householders can cut their energy and water use by nearly half with little or no cost, as well as generate large long-term savings, ranging from $1,000 to $10,000 annually (chapter 5). Moreover, the land associated with homes (typically yards) can be made more habitable for wildlife and less damaging for ecosystems. Because houses have largely been exempt from the rush toward efficiency in the industrial and commercial sectors, homeowners have more easy, low-cost options for reducing their energy, water, and land use than any other sector. The strategic development of new homes and a careful retrofitting of older ones can directly contribute to climate-change mitigation and lead to more sustainable communities.

Despite the incredible opportunities for individual householders to cash in on making their homes more sustainable, cities may be the key to turning the tide toward sustainable household dynamics (chapter 6). Cities have more motivation than individual householders, states, or nations to address the housing bomb. First, those in cities know that the sprawl cycle makes today's prospering community tomorrow's decaying neighborhood, unless they can provide key amenities not available in the next ring of suburbia. A green infrastructure for households that capitalizes on existing infrastructures and the current population density (e.g., mixed-use development, or light-rail systems) gives existing neighborhoods one of their few advantages over the ones in the process of being built farther afield.

A household-based green infrastructure can also contribute to sustainable communities in less obvious ways. Globalization is reducing the reliability of job markets. Increasing global competitiveness means that jobs tend to flow toward the most-productive and least-expensive workforce. At the same time, globalization has allowed a small, highly educated sector of the global population to choose where they live, and their talents therefore concentrate where global revenue is located. Groups of these highly sought-after people already have begun clustering in areas with environmental amenities (chapter 6). Communities that develop options for green homes and infrastructures will have an advantage when competing for these workers, because people who know they have other options rarely choose communities plagued by smog and toxins.

A household-based energy infrastructure also encourages public commitment to education, an essential ingredient for success in global markets. People who understand where their energy comes from also understand how important it is to produce and maintain local talent capable of managing those energy supplies. It is difficult for individual householders to benefit from solar and wind power, but cities can, because they are more readily able to afford the outlays associated with large (and thus cost-efficient) systems. Community investment in green infrastructures, ranging from energy production to pedestrian-friendly streets, creates strong ties to the community. Unlike commonalities formed through shared debt, however, these positive bonds between residents and the places where they live can create profit (for both parties) and community vitality.

Cities across many nations have instigated pedestrian and bicycle-friendly infrastructures that connect homes with workplaces and shopping areas, and most cities have done so for less money that it costs to pave three miles of urban interstate highway (chapter 6). Almost across the board, these municipalities have seen 10%–20% increases in the number of people walking and cycling while reducing the number—not only the rate, but also the number— of accidents and fatalities among pedestrians and bicyclists. This shift makes a city's populace healthier, wealthier, and more energy efficient. Similar gains have been made in water efficiency. According to the Southern Nevada Water Authority (www.snwa.com), Las Vegas's program of paying residents to replace turfgrass lawns with xeric landscaping (current prices are between $1 and $1.50 per square foot) was responsible for eliminating over 150 million square feet of turfgrass by 2011. Given that Las Vegas draws its water from the Colorado River, which regularly dries up before reaching the Gulf of California, this program has the potential to breathe life back into the artificially parched downriver landscape. If similar programs were adopted more widely, entire watersheds could be saved by the households now owning water rights that were previously allocated to agriculture. Cities around the country have instituted comprehensive plans that promote density near locales with public transportation; mayors are scrambling for light-rail systems; and planning efforts associated with smart growth and new urbanism are being rapidly adopted by cities hoping to avoid being the next step in the sprawl cycle, where urban decay spreads and the tax bases move progressively farther and farther from city centers.

The efforts by cities to promote safe alternatives to driving and connected streets and neighborhoods may also help counter concerns about humanity's isolation from nature. As the United States moved from urbanization to suburbanization, the primary cause of this isolation shifted from the cities themselves to personal motor vehicles. The widespread separation of society from nature was a lengthy process, beginning with the conversion of hunter-gatherer cultures to agrarian societies, and culminating with technological advances that are freeing household-location decisions from the constraints of being adjacent to places of employment. These developments fracture the relationships between humans and nature. Without them, people feel no tangible connections between the decisions they make and their impacts on the various elements that constitute nature. Natural environments in the vicinity of homes have traditionally provided a treatment for the ills of nature-deficit disorder.[8] Traditional suburban homes are among the last places where humans interact with nature. They are where people plant gardens, feed birds, keep pets, plant trees, and fertilize grass. National complete-streets and smart-growth movements cannot entirely undo the ecological damage wrought by sprawl and low-density development, but by getting people out of their vehicles and back into their yards, streets, and neighborhoods, they provide an unprecedented opportunity for us to connect with nature in a tangible way.

States and nations can promote their own competiveness and sustainability by supporting householders and cities in their struggle to defuse the housing bomb (chapter 7). Over the next decade in the United States, the individual states and the federal government face projected health-care costs reaching $1 trillion annually, solely from obesity. These costs will hamstring economic growth, and the least-expensive way to address them is to make human-powered transportation safe and feasible for more citizens. Similarly, peak oil makes the United States as a whole, and every state in the nation, more vulnerable to rising oil prices than any other country in the world, with the possible exceptions of Russia and some Middle Eastern nations.

State and federal support for householder and local efforts to address the housing bomb is essential for several reasons that go beyond addressing health-care costs and becoming resilient in the face of peak oil (chapter 7). First, the legacy of policies promoting sprawling development, inefficient low-quality homes, artificially high levels of home ownership, and inadequate information about the benefits of sustainable housing means that without governmental intervention, markets will produce suboptimal outcomes for most people. Second, current fiscal and policy frameworks associated with financing and insuring homes, and providing transportation infrastructures to and from homes, is controlled by nations and states (chapters 2 and 6). Until that power is shifted to cities, they lack critical resources for addressing the housing bomb. Third, as demonstrated by the Portland, Oregon / Washington and Tahoe, Nevada / California regions, many urban planning issues cross state boundaries and require more political power to address than cities can muster.

Ironically, urban slums in the world's largest cities offer some of the most promising possibilities for defusing the housing bomb (chapter 7). Because these slums have not sprawled across the landscape, they do not need to become more dense. The extreme poverty of their inhabitants has led to walkable distances between homes and many needed services and products. Building a smart-energy grid is one place where urban slums are poised for improving our approach to household dynamics. Even before the 2010 hurricane that destroyed Haiti's infrastructure, Port-au-Prince was plagued with energy problems.[9] Most residents had no regular access to electricity. A Colorado-based firm, Green Energy Corp, has developed a smart-grid plan for Haiti, and this firm is now working with a suite of partners (including the U.S. Department of Defense, the U.S. Department of Energy, the United Nations' Environment Programme, the Clinton Global Initiative, Duke Energy, and EarthSpark International) to develop an energy grid (using sun, wind, water, and biomass) that will deliver electricity to Haitians. By focusing their efforts on a distribution model that links small-scale renewable sources together closer to the load (sites of use), they have reduced transmission costs. And it doesn't hurt that the small wind and solar farms have created new jobs for Haitians. Programs such as this offer great possibilities for contributing to the already-existing informal economies that thrive in these completely mixed-use communities.

The problems of house addiction and homevoting (Introduction and chapter 6) may actually hold several keys to finally capitalizing on all the opportunities associated with defusing the housing bomb. Houses provide a powerful addiction. The good news is that there is simply no reason why that addiction must be limited to the number of square feet in one's home and to lot size. Sustainability efforts can capitalize on the emerging passion for comparing and ranking our homes by providing public information on items such as energy use and water consumption. This may seem like a stretch to some, but the same phenomenon has worked exceptionally well for hybrid cars. When they were introduced they were universally panned, but once they became a status symbol, and a means of demonstrating neighborhood rank, they rapidly gained a large market share.

Similarly, homevoters are painted as being at the root of sprawl, neighborhood segregation, and a reluctance to adopt principles of smart growth, but these tendencies can be overcome (chapter 6). Moreover, the desperation to protect property values rooted in overinvestment in real estate can be turned to more positive ends. Alliances between homeowners hoping to increase their property values and NGOs such as land trusts have demonstrated the ability to create built-out scenarios (contexts where additional development is no longer viable) without building on open space, thereby pushing development back into urban centers (chapters 3 and 6).[10] Similarly, if homevoters are educated about the sprawl cycle, they would presumably fight to stop it before their neighborhoods begin to decline.[11] The sprawl cycle—where the non-poor abandon existing communities in favor of cheap rural subdivisions with lower taxes—should terrify most homevoters.

Strategies for defusing the housing bomb include feel-good suggestions, such as renovating urban cores, making suburbia more environmentally friendly (e.g., conservation subdivisions), and redirecting house addiction into the green consumption of household-related amenities. Unfortunately, any real solution also will include less-popular changes that would require people to pay the full social and environmental costs for choosing to live far from centers of employment, educational institutions, and retail establishments. Thus "houseaholism" will win out over environmentalism and altruism, unless there are enforced regulations or market mechanisms that make householders pay the true costs of their housing. It would be nice if the most vocal critics of sprawl got roommates and moved to downtown apartments, but that scenario will not be sufficient, even in the unlikely event it actually materializes. Communities must ensure that householders are responsible for the overall costs of their choices and actions, including sprawl, standalone homes, huge home sizes, and homes shared with fewer people. In these efforts, communities could benefit from capitalizing on the aversion to entitlements espoused by the political Right. Moving to rural areas should not entitle someone to dodge the taxes needed to pay for social services in their community, state, or nation. People living in suburbia should not be entitled to subsidized roads, sewers, fire protection, police protection, or schools. For every mile farther from a city center that someone lives, those services cost more to provide, and they typically cause more environmental damage (chapter 3). Accordingly, various forms of payment (including taxes) to cover these costs should go up as distances increase between residents and the services they need. Similarly, taking out a huge home loan should not entitle someone to tax subsidies (income-tax deductions for the interest charged on mortgage payments) that are denied to people who rent their housing. Unless a fair distribution of costs can be implemented, sustainable household dynamics will not be achieved.

Framing sustainability in terms of household impacts is another way to help promote a fair distribution of opportunities, costs, and rewards in society. The burden of conservation would then shift from the backs of individuals with the lowest educational levels, little income, and less concern about the environmental impacts of their houses and onto people with the highest educational levels, more personal wealth, and greater concern about the effects their houses have on the environment.[12] Conservation efforts driven from a household perspective would push people who love nature, have substantial incomes, and have an advanced level of education to sacrifice what they *want* (e.g., a home in the countryside, on a river, on a mountainside, or on fragile desert soils) before expecting the poor or individuals with lower levels of education to sacrifice what they *need* (e.g., heating, health care, or a better education for their children) in the name of sustainability.

For all these reasons and more, a household-dynamics perspective can help us move beyond the apocalyptic rhetoric of old-school environmentalism and start remodeling our world. Restructuring that saves energy, water, and clean air at the household level offers people multiple opportunities to become part of the solution, rather than limiting us to being part of the problem. Sustainability requires society to move beyond regrets, anger, and guilt regarding past relationships between humans and nature. A household-dynamics approach offers clear guidance for that move, since it shifts our efforts in the direction of building a more sustainable future. In today's globalized world, the fact that a household-dynamics approach is not directly tied to specific economic or political dogmas is another advantage. It does not require us to advocate for capitalism, socialism, or any other economic system. It can work in strong democracies and totalitarian political regimes.

It does not care whether a given remodeling project is supported by public or private funds, or by a combination of the two. It only cares that the financing is appropriate for the cultural setting.

Robert Fishman has predicted that the housing bomb may be defused in part by what he labels the fifth migration, where affluent households return to cities and first-tier (close-in) suburbs.[13] He suggests that downtown reurbanism, immigrant reurbanism, black reurbanism, and white middle-class reurbanism will drive this migration. It would certainly be more sustainable than the fourth migration (a diaspora from the cities into the suburbs). Whether the fifth migration will happen remains to be seen, but all the evidence points toward increasing demands for small lots, high density, mixed uses, and walkability (chapter 6). Builders will respond to substantial demands for detached houses on small lots and multiunit homes, and markets will respond to plummeting purchases of detached houses on large lots.[14] The interacting forces of a declining preference for detached homes on spacious individual acreage and baby boomers trying to sell their high-priced houses to members of less-privileged generations may gut the demand for McMansions.[15]

These shifts will help address the housing bomb, but they will be extremely painful for the elderly and the poor (chapter 6). These groups may be trapped in second-tier, vehicle-dependent suburbs with declining property values unless rapid and strategic actions are taken to ensure social justice. These changes include building affordable housing in first-tier suburbs, using in-fill projects (building on vacant property) in first-tier suburbs to achieve sufficient density for public transportation, and developing pedestrian and bicycle-friendly infrastructures. The latter two efforts can reduce the footprints of metro areas and first-tier suburbs by offering replacements for personal automotive transport, and they would allow residents of second-tier suburbs to safely reach the nearest forms of public transit and nearby retail establishments. These changes carry the added benefits of reducing skyrocketing obesity rates and vulnerability to peak oil (chapters 6 and 7).

There are good reasons why people choose to remodel their homes, rather than tearing them down and starting from scratch. We think the same reasons apply to society. The basic structures are worth saving; they just need to be updated. The possibilities range from the straightforward and relatively simple fixes for existing homes, neighborhoods, and cities that are described in this book to futuristic designs for entirely new ways of living. Emphasizing the critical role households can play in sustainability carries the added benefit of reengaging humans with nature. Many people realize that society's current trajectory is not sustainable. Climate change has provided a dramatic reminder that even today's children, not just generations to come, will suffer for our careless use of the planet. Since we are not prepared to tear down the whole edifice, it's time to start remodeling. Remodeling our homes offers, both literally and figuratively, the key to a sustainable future on our larger home, Earth.

Introduction

1. Borlaug, N. E. Ending world hunger: The promise of biotechnology and the threat of antiscience zealotry. *Plant Physiology* **124**, 487–490 (2000).

2. Raudsepp-Hearne, C. et al. Untangling the environmentalist's paradox: Why is human well-being increasing as ecosystem services degrade? *Bioscience* **60**, 576–589 (2010), doi:10.1525/bio.2010.60.8.4.

3. Sheeran, J. How to end hunger. *Washington Quarterly* **33**, 3–16 (2010).

4. Morgan, S. P. & King, R. B. Why have children in the 21st century? Biological predisposition, social coercion, rational choice. *European Journal of Population / Revue europénne de Démographie* **17**, 3–20 (2001).

5. Bulatao, R. A. Values and disvalues of children in successive childbearing decisions. *Demography* **18**, 1–25 (1981).

6. Bulatao, R. A. Values and disvalues of children.

7. United Nations, Population Division. *World Population Prospects: The 2008 Revision*. United Nations, 2008.

8. McGinn, D. F. *House Lust: America's Obsession with Our Homes*. Random House, 2008.

9. Liu, J. G., Daily, G. C., Ehrlich, P. R. & Luck, G. W. Effects of household dynamics on resource consumption and biodiversity. *Nature* **421**, 530–533 (2003).

10. Keilman, N. Biodiversity: The threat of small households. *Nature* **421**, 489–490 (2003).

11. Peterson, M. N., Peterson, M. J., Peterson, T. R. & Liu, J. G. A household perspective for biodiversity conservation. *Journal of Wildlife Management* **71**, 1243–1248 (2007), doi:10.2193/2006-207.

12. Czech, B., Krausman, P. R. & Devers, P. K. Economic associations among causes of species endangerment in the United States. *Bioscience* **50**, 593–601 (2000).

13. Peterson, M. N., Chen, X. D. & Liu, J. G. Household location choices: Implications for biodiversity conservation. *Conservation Biology* **22**, 912–921 (2008), doi:10.1111/j.1523-1739.2008.00929.x.

14. Radeloff, V. C. et al. Housing growth in and near United States protected areas limits their conservation value. *Proceedings of the National Academy of Sciences of the United States of America* **107**, 940–945 (2010), doi:10.1073/pnas.0911131107.

15. McGinn, D. F. *House Lust*.

16. McGinn, D. F. *House Lust.*

17. Leopold, A. *A Sand County Almanac and Sketches Here and There*, 48. Oxford University Press, 1949.

18. Kahn, M. E. & Morris, E. A. Walking the walk: The association between community environmentalism and green travel behavior. *Journal of the American Planning Association* 75, 389–405 (2009).

19. Worster, D. *Nature's Economy: A History of Ecological Ideas.* Cambridge University Press, 1994.

20. Leopold, A. *A Sand County Almanac.*

21. Lovelock, J. *The Ages of Gaia: A Biography of Our Living Planet.* Norton, 1988.

22. Worster, D. *Nature's Economy*, 382.

23. Peterson, M. N., Chen, X. D. & Liu, J. G. Household location choices.

24. On attention deficit disorder, Faber Taylor, A. & Kuo, F. E. Children with attention deficits concentrate better after walk in the park. *Journal of Attention Disorders* 12, 402 (2009). On obesity, Potwarka, L. R., Kaczynski, A. T. & Flack, A. L. Places to play: Association of park space and facilities with healthy weight status among children. *Journal of Community Health* 33, 344–350 (2008). On asthma, Lovasi, G. S., Quinn, J., Neckerman, K., Perzanowski, M. & Rundle, A. Children living in areas with more street trees have lower prevalence of asthma. *Journal of Epidemiology and Community Health* 62, 647 (2008). On myopia, Rose, K. A. et al. Outdoor activity reduces the prevalence of myopia in children. *Ophthalmology* 115, 1279–1285 (2008). On stress, Wells, N. M. & Evans, G. W. Nearby nature. *Environment and Behavior* 35, 311 (2003). On mental health, Pretty, J., Peacock, J., Sellens, M. & Griffin, M. The mental and physical health outcomes of green exercise. *International Journal of Environmental Health Research* 15, 319–337 (2005).

25. Papas, M. A. et al. The built environment and obesity. *Epidemiologic Reviews* 29, 129 (2007). Nelson, M. C., Gordon-Larsen, P., Song, Y. & Popkin, B. M. Built and social environments: Associations with adolescent overweight and activity. *American Journal of Preventive Medicine* 31, 109–117 (2006).

26. Nelson, M. C. et al., Built and social environments.

27. Louv, R. *Last Child in the Woods: Saving Our Children from Nature Deficit Disorder.* Algonquin Books, 2005.

28. Plato. *The Republic.* Heritage Press, 1944.

29. Latour, B. *Politics of Nature: How to Bring the Sciences into Democracy*, 11. Harvard University Press, 2004.

30. Haraway, D. *Modest_Witness@Second_Millennium:FemaleMan©_Meets_Onco Mouse™; Feminism and Technoscience.* Routledge, 1997.

31. Louv, R. *Last Child in the Woods.* Spurlock, M. et al. *Supersize Me.* Video recording. Presented by Roadside Attractions, Samuel Goldwyn Films, and Showtime Films. Produced by M. Spurlock and The Con. Hart Sharp Video, 2004.

32. Peterson, M. N., Peterson, T. R., Peterson, M. J., Lopez, R. R. & Silvy, N. J. Cultural conflict and the endangered Florida Key deer. *Journal of Wildlife Management* 66, 947–968 (2002).

33. Hardin, G. The tragedy of the commons. *Science* 162, 1243–1248 (1968).

34. Owen, D. *Green Metropolis: Why Living Smaller, Living Closer, and Driving Less Are the Keys to Sustainability.* Penguin, 2009.

35. United Nations. *The Millennium Development Goals Report, 2010*. United Nations, 2010.

36. Dowie, M. Conservation refugees. *Orion* **November/December**, 16–27 (2005). Radeloff, V. C. et al. Housing growth.

37. Liu, J. G. et al. Effects of household dynamics. Radeloff, V. C. et al. Housing growth.

38. Radeloff, V. C. et al. Housing growth.

39. National Association of Realtors, Research Division. *2010 NAR Investment and Vacation Home Buyers Survey*. National Association of Realtors, 2010.

40. National Association of Realtors. *2010 NAR Survey*.

Chapter 1 · Household Dynamics and Their Contribution to the Housing Bomb

1. Wolman, M. G. U.S. case studies: An introduction, in *Growing Populations, Changing Landscapes: Studies from India, China, and the United States*, 233–236. National Academy Press, 2001.

2. Wolman, M. G. U.S. case studies.

3. de Sherbinin, A. Water and population dynamics: Local approaches to a global challenge, in *Water and Population Dynamics: Case Studies and Policy Implications; Report of a Workshop, October 1996, Montreal, Canada* (ed. V. Dompka & A. de Sherbinin, with L. Bromley), 9–22. American Association for the Advancement of Science, 1998. Thompson, K. & Jones, A. Human population density and prediction of local plant extinction in Britain. *Conservation Biology* **13**, 185–189 (1999).

4. Klinenberg, E. *Going Solo: The Extraordinary Rise and Surprising Appeal of Living Alone*. Penguin Press, 2012.

5. Klinenberg, E. *Going Solo*.

6. Diamond, J. M. *Collapse: How Societies Choose to Fail or Succeed*. Viking, 2005. Liu, J. G. & Diamond, J. China's environment in a globalizing world. *Nature* **435**, 1179–1186 (2005).

7. Liu, J. G., Daily, G. C., Ehrlich, P. R. & Luck, G. W. Effects of household dynamics on resource consumption and biodiversity. *Nature* **421**, 530–533 (2003).

8. For CO_2 emissions, MacKellar, F. L., Lutz, W., Prinz, C. & Goujon, A. Population, households, and CO_2 emissions. *Population and Development Review* **21**, 849–865 (1995). For fuelwood consumption, Clinecole, R. A., Main, H. A. C. & Nichol, J. E. On fuelwood consumption, population-dynamics and deforestation in Africa. *World Development* **18**, 513–527 (1990). Hosier, R. Household energy-consumption in rural Kenya. *Ambio* **14**, 225–227 (1985). Kaul, S. & Qian, L. Rural household energy use in China. *Energy* **17**, 405–411 (1992). For vehicle use, Liddle, B. Demographic dynamics and per capita environmental impact: Using panel regressions and household decompositions to examine population and transport. *Population and Environment* **26**, 23–39 (2004). For species endangerment, Peterson, M. N., Peterson, M. J., Peterson, T. R. & Liu, J. G. A household perspective for biodiversity conservation. *Journal of Wildlife Management* **71**, 1243–1248 (2007), doi:10.2193/2006-207.

9. MacKellar, F. L. et al. Population, households, and CO_2 emissions.

10. Liu, J. G. et al. Effects of household dynamics.

11. Pachauri, R. K. & Intergovernmental Panel on Climate Change. *Climate Change 2007: Synthesis Report; A Report of the Intergovernmental Panel on Climate Change.* Intergovernmental Panel on Climate Change, 2008. Dietz, T., Gardner, G. T., Gilligan, J., Stern, P. C. & Vandenbergh, M. P. Household actions can provide a behavioral wedge to rapidly reduce US carbon emissions. *Proceedings of the National Academy of Sciences of the United States of America* **106**, 18452–18456 (2009).

12. Bin, S. & Dowlatabadi, H. Consumer lifestyle approach to US energy use and the related CO_2 emissions. *Energy Policy* **33**, 197 (2005).

13. Dietz, T. et al. Household actions.

14. Pachauri, S. *An Energy Analysis of Household Consumption: Changing Patterns of Direct and Indirect Use in India.* Springer, 2007, http://dx.doi.org/10.1007/978-1-4020-5712-0.

15. O'Neill, B. C. et al. Global demographic trends and future carbon emissions. *Proceedings of the National Academy of Sciences of the United States of America* **107**, 17521–17526 (2010).

16. Bongaarts, J. Household size and composition in the developing world in the 1990s. *Population Studies: A Journal of Demography* **55**, 263–279 (2001).

17. Frankel, D. & Webb, J. M. Population, households, and ceramic consumption in a prehistoric Cypriot village. *Journal of Field Archaeology* **28**, 115–129 (2001). Burch, T. K. Size and structure of families—comparative analysis of census data. *American Sociological Review* **32**, 347–363 (1967). Liao, T. F. T. Were past Chinese families complex? Household structures during the Tang Dynasty, 618–907 AD. *Continuity and Change* **16**, 331–355 (2001). Laslett, P. A. *Household and Family in Past Time.* Cambridge, 1972.

18. Liu, J. G. et al. Effects of household dynamics.

19. United Nations, Statistics Division. *Millennium Development Indicators: World and Regional Groupings.* United Nations, 2003.

20. This trend reflects those found for demographic transitions in population growth.

21. Donnelly, J. S. *The Great Irish Potato Famine.* Sutton, 2001.

22. United Nations, Human Settlements Programme. *Enhancing Urban Safety and Security: Global Report on Human Settlements 2007.* Earthscan, 2007.

23. U.S. Census. Highlights of Annual 2012 Characteristics of New Housing, 2012, www.census.gov/construction/chars/highlights.html, accessed June 1, 2013.

24. Allen, S. C., Moorman, C. E., Peterson, M. N., Hess, G. R. & Moore, S. E. Overcoming socio-economic barriers to conservation subdivisions: A case-study of four successful communities. *Landscape and Urban Planning* **106**, 244–252 (2012).

25. Liu, J. et al. Effects of household dynamics. Lenzen, M. & Murray, S. A. A modified ecological footprint method and its application to Australia. *Ecological Economics: The Journal of the International Society for Ecological Economics* **37**, 229 (2001). Yousif, H. M. Population, biomass and the environment in central Sudan. *International Journal of Sustainable Development & World Ecology* **2**, 54–69 (1995).

26. World Bank. *World Development Indicators.* World Bank, 2012.

27. Keilman, N. Biodiversity: The threat of small households. *Nature* **421**, 489–490 (2003).

28. United Nations, Population Division. *Living Arrangements of Older Persons Around the World*. United Nations, 2005.

29. Klinenberg, E. *Going Solo*.

30. Klinenberg, E. *Going Solo*.

31. Rosenfeld, M. J. *The Age of Independence: Interracial Unions, Same-Sex Unions, and the Changing American Family*. Harvard University Press, 2007.

32. Yu, E. & Liu, J. Environmental impacts of divorce. *Proceedings of the National Academy of Sciences of the United States of America* **104**, 20629–20634 (2007).

33. Furstenberg, F. F. & Cherlin, A. J. *Divided Families: What Happens to Children when Parents Part*, vol. 1. Harvard University Press, 1991.

34. Yu, E. & Liu, J. Environmental impacts of divorce.

35. Yu, E. & Liu, J. Environmental impacts of divorce.

36. Thornton, A., Axinn, W. G. & Xie, Y. *Marriage and Cohabitation*. University of Chicago Press, 2007.

37. Hellerstein, J. K. & Morrill, M. S. Booms, busts, and divorce. *B.E. Journal of Economic Analysis and Policy* **11**, article 54 (2011).

38. National Association of Home Builders. *Housing Facts, Figures and Trends*.

39. Liu, J. & Diamond, J. China's environment.

40. Ekland-Olson, S., Barrick, D. M. & Cohen, L. E. Prison overcrowding and disciplinary problems: An analysis of the Texas prison system. *Journal of Applied Behavioral Science* **19**, 163–176 (1983).

41. Soule, D. C. Defining and managing sprawl, in *Urban Sprawl: A Comprehensive Reference Guide* (ed. D. C. Soule), 3–11. Greenwood Press, 2006.

42. Ewing, R., Pendall, R. & Chen, D. *Measuring Sprawl and Its Impact: The Character and Consequences of Metropolitan Expansion*. Smart Growth America, 2002, www.smartgrowthamerica.org/research/measuring-sprawl-and-its-impact/, accessed January 3, 2013.

43. Owen, D. *Green Metropolis: Why Living Smaller, Living Closer, and Driving Less Are the Keys to Sustainability*. Penguin, 2009. Gonzalez, G. A. *Urban Sprawl, Global Warming, and the Empire of Capital*. State University of New York Press, 2009.

44. Hirschhorn, J. S. Environment, quality of life, and urban growth in the new economy. *Environmental Quality Management* **10**, 1–8 (2001).

45. Gardner, S. The impact of sprawl on the environment and human health, in *Urban Sprawl* (ed. D. C. Soule), 240–259.

46. Ewing, R., Pendall, R. & Chen, D. *Measuring Sprawl*. Hirschhorn, J. S. Environment, quality of life, and urban growth.

47. Milesi, C., Elvidge, C. D., Dietz, J. B., Tuttle, B. T. & Nemani, R. B. Mapping and modeling the biogeochemical cycling of turfgrasses in the United States. *Environmental Management* **36**, 426–438 (2005).

48. Robbins, P. & Sharp, J. T. Producing and consuming chemicals: The moral economy of the American lawn, in *Urban Ecology: An International Perspective on the Interactions between Humans and Nature* (ed. J. M. Marzluff et al.), 180–206. Springer, 2008.

49. Milesi, C. et al. Mapping and modeling biogeochemical cycling. Kaye, J. P., Groffman, P. M., Grimm, N. B., Baker, L. A. & Pouyat, R. V. A distinct urban biogeochemistry? *Trends in Ecology & Evolution* **21**, 192–199 (2006).

50. Gardner, S. The impact of sprawl, in *Urban Sprawl* (ed. D. C. Soule), 240–259. Bijoor, N. S., Czimczik, C. I., Pataki, D. E. & Billings, S. A. Effects of temperature and fertilization on nitrogen cycling and community composition of an urban lawn. *Global Change Biology* **14**, 2119–2131 (2008).

51. Gardner, S. The impact of sprawl, in *Urban Sprawl* (ed. D. C. Soule), 240–259.

52. Gardner, S. The impact of sprawl, in *Urban Sprawl* (ed. D. C. Soule), 240–259. Karathodorou, N., Graham, D. J. & Noland, R. B. Estimating the effect of urban density on fuel demand. *Energy Economics* **32**, 86–92 (2010). Newman, P. W. G. & Kenworthy, J. R. Gasoline consumption and cities. *Journal of the American Planning Association* **55**, 24–37 (1989).

53. Gardner, S. The impact of sprawl, in *Urban Sprawl* (ed. D. C. Soule), 240–259.

54. Ewing, R., Kostyack, J., Chen, D., Stein, B. & Ernst, M. *Endangered by Sprawl: How Runaway Development Threatens America's Wildlife.* National Wildlife Federation, Smart Growth America, and Nature Serve, 2005.

55. Gardner, S. The impact of sprawl, in *Urban Sprawl* (ed. D. C. Soule), 240–259.

56. Yim, S. H. L. & Barrett, S. R. H. Public health impacts of combustion emissions in the United Kingdom. *Environmental Science & Technology* **46**, 4291–4296 (2012).

57. Gardner, S. The impact of sprawl, in *Urban Sprawl* (ed. D. C. Soule), 240–259.

58. Ewing, R., Schmid, T., Killingsworth, R., Zlot, A. & Raudenbush, S. Relationship between urban sprawl and physical activity, obesity, and morbidity, in *Urban Ecology* (ed. J. M. Marzluff et al.), 567–582. Springer, 2008.

59. Davis, S. C., Diegel, S. W. & Boundy, E. G. *Transportation Energy Data Book: Edition 30.* Oak Ridge National Laboratory, 2011.

60. Gardner, S. The impact of sprawl, in *Urban Sprawl* (ed. D. C. Soule), 240–259.

61. Rusk, D. Social framework: Sprawl, race, and concentrated poverty—changing the "rules of the game," in *Urban Sprawl* (ed. D. C. Soule), 90–102.

62. Hirsch, R., Bezdek, R. & Wendling, R. *Peaking of World Oil Production: Impacts, Mitigation, & Risk Management.* Report presented to the U.S. Department of Energy, 2005.

63. Hirsch, R., Bezdek, R. & Wendling, R. *Peaking of World Oil Production.*

64. Brown, L. R. *Plan B 3.0: Mobilizing to Save Civilization.* W. W. Norton, 2008.

65. Gonzalez, G. A. *Urban Sprawl.*

66. Peterson, M. N. & Liu, J. G. Property rights and landscape planning in the Intermountain West: The Teton Valley case. *Landscape and Urban Planning* **86**, 126–133 (2008), doi:10.1016/j.landurbplan.2008.01.003.

67. Allen, S. C. et al. Overcoming socio-economic barriers.

68. Tranter, P. *Effective Speeds: Car Costs Are Slowing Us Down.* Australian Greenhouse Office, 2004, www.environment.gov.au/archive/settlements/transport/publications/effectivespeeds.html, accessed January 11, 2013. Burrington, S. H. & Conservation Law Foundation. *Road Kill: How Solo Driving Runs Down the Economy.* Conservation Law Foundation, 1994.

69. Brueckner, J. K. Urban sprawl: Diagnosis and remedies. *International Regional Science Review* **23**, 160–171 (2000).

70. Gonzalez, G. A. *Urban Sprawl.*

71. Weiss, M. A. *The Rise of the Community Builders: The American Real Estate*

Industry and Urban Land Planning. Columbia University Press, 1987. Warner, S. B. *Streetcar Suburbs: The Process of Growth in Boston, 1870–1900.* Harvard University Press, 1978.

72. Wickersham, J. Legal framework: The laws of sprawl and the laws of smart growth, in *Urban Sprawl* (ed. D. C. Soule), 26–60.

73. Gonzalez, G. A. *Urban Sprawl.*

74. Gonzalez, G. A. *Urban Sprawl.*

75. Hirsch, R. L. Mitigation of maximum world oil production: Shortage scenarios. *Energy Policy* **36**, 881–889 (2008).

76. Liu, J. & Taylor, W. W. *Integrating Landscape Ecology into Natural Resource Management.* Cambridge University Press, 2002.

77. Friesen, L. E., Eagles, P. F. J. & Mackay, R. J. Effects of residential development on forest-dwelling neotropical migrant songbirds. *Conservation Biology* **9**, 1408–1414 (1995). Kluza, D. A., Griffin, C. R. & Degraaf, R. M. Housing developments in rural New England: Effects on forest birds. *Animal Conservation* **3**, 15–26 (2000). Nilon, C. H., Long, C. N. & Zipperer, W. C. Effects of wildland development on forest bird communities. *Landscape and Urban Planning* **32**, 81 (1995).

78. Gardner-Outlaw, T. & Engelman, R. *Forest Futures: Population, Consumption, and Wood Resources.* Population Action International, 1999.

79. Liu, J. & Diamond, J. China's environment.

80. Hawbaker, T. J., Radeloff, V. C., Clayton, M. K., Hammer, R. B. & Gonzalez-Abraham, C. E. Road development, housing growth, and landscape fragmentation in northern Wisconsin: 1937–1999. *Ecological Applications: A Publication of the Ecological Society of America* **16**, 1222–1237 (2006).

81. Smith, C. M. & Wachob, D. G. Trends associated with residential development in riparian breeding bird habitat along the Snake River in Jackson Hole, WY, USA: Implications for conservation planning. *Biological Conservation* **128**, 431–446 (2006).

82. Smith, C. M. & Wachob, D. G. Trends associated with residential development. Gonzalez-Abraham, C. E. et al. Patterns of houses and habitat loss from 1937 to 1999 in northern Wisconsin, USA. *Ecological Applications: A Publication of the Ecological Society of America* **17**, 2011–2023 (2007).

83. Peralta, G., Fenoglio, M. S. & Salvo, A. Physical barriers and corridors in urban habitats affect colonisation and parasitism rates of a specialist leaf miner. *Ecological Entomology* **36**, 673–679 (2011).

84. Beier, P. Determining minimum habitat areas and habitat corridors for cougars. *Conservation Biology* **7**, 94–108 (1993).

85. Gavier-Pizarro, G. I., Radeloff, V. C., Stewart, S. I., Huebner, C. D. & Keuler, N. S. Rural housing is related to plant invasions in forests of southern Wisconsin, USA. *Landscape Ecology* **25**, 1505–1518 (2010).

86. Klinenberg, E. *Going Solo: The Extraordinary Rise and Surprising Appeal of Living Alone.* Penguin Press, 2012.

87. Leinberger, C. B. & Alfonzo, M. *Walk This Way: The Economic Promise of Walkable Places in Metropolitan Washington, D.C.* Brookings Institution, 2012.

88. Leinberger, C. B. Now coveted: A walkable, convenient place. *New York Times*, May 27 (2012), SR6.

89. Nelson, A. C. Leadership in a new era. *Journal of the American Planning Association* 72, 393–407 (2006).

90. Leinberger, C. B. & Alfonzo, M. *Walk This Way.*

Chapter 2 · How Home Ownership Both Emancipates and Enslaves Us

1. Harvey, D. The right to the city. *New Left Review* 53, 23–40 (2008).

2. Ronald, R. Home ownership, ideology and diversity: Re-evaluating concepts of housing ideology in the case of Japan. *Housing, Theory & Society* 21, 49–64 (2004).

3. Jackson, B. Revisionism reconsidered: "Property-owning democracy" and egalitarian strategy in post-war Britain. *Twentieth-Century British History* 16, 416–440 (2005).

4. United Kingdom. *Housing Act 1980* (chapter 51).

5. Rawnsley, A. Thatcher's dream becomes a nightmare for a jilted generation. *Observer*, June 4 (2011).

6. Harvey, D. The right to the city.

7. Australian Housing and Urban Research Institute. *What Can Australia Learn from International Trends in Housing and Policy Responses?* AHURI Research & Policy Bulletin 97, 2008.

8. Lands, L. Be a patriot, buy a home: Re-imagining home owners and home ownership in early 20th-century Atlanta. *Journal of Social History* 41, 943–965, 1102–1103 (2008). Hanan, J. S. Home is where the capital is: The culture of real estate in an era of control societies. *Communication & Critical/Cultural Studies* 7, 176–201 (2010).

9. Lands, L. Be a patriot, 944.

10. Lands, L. Be a patriot.

11. Lands, L. Be a patriot, 950.

12. Lands, L. Be a patriot, 944.

13. Lands, L. Be a patriot, 949.

14. Lands, L. Be a patriot, 948.

15. Lands, L. Be a patriot, 949.

16. The National Association of Real Estate Boards was founded in 1908 as the National Association of Real Estate Exchanges, and it later changed its name to the National Association of Real Estate Boards. Its current name, the National Association of Realtors (now using the acronym NAR) was adopted in 1974. It is the largest trade association associated with real estate in the United States. The National Association of Real Estate Brokers (now using the acronym NAREB) was formed by black real estate professionals in 1947. They have focused on ensuring equal housing opportunities for all Americans.

17. Lands, L. Be a patriot, 952.

18. Lands, L. Be a patriot, 943.

19. Marciano, R., Goldberg, D. & Hou, C.-Y. *T-Races: A Testbed for the Redlining Archives of California's Exclusionary Spaces*, 2012. University of California Humanities Research Institute, http://salt.unc.edu/T-RACES/, accessed January 28, 2013.

20. Marciano, R., Goldberg, D. & Hou, C.-Y. *T-Races.*

21. Marciano, R., Goldberg, D. & Hou, C.-Y. *T-Races.*

22. U.S. Federal Housing Administration. *Underwriting Manual*, Section 310. U.S. Government Printing Office, 1935.

23. Article 34, quoted in Stern, S. M. Residential protectionism and the legal mythology of home. *Michigan Law Review* **107**, 1093–1144 (2010).

24. Marciano, R., Goldberg, D. & Hou, C.-Y. *T-Races.*

25. Stern, S. M. Residential protectionism, 1094.

26. Stern, S. M. Residential protectionism, 1094.

27. Stern, S. M. Residential protectionism, 1095.

28. Radin, M. J. Property and personhood. *Stanford Law Review* **34**, 957–959 (1982).

29. Krugman, P. Home not-so-sweet home. *New York Times,* June 23 (2008), 21A.

30. Stern, S. M. Residential protectionism.

31. Hines, R. M. Credit markets, exemptions, and households with nothing to exempt. *Theoretical Inquiries in Law* **493**, 512–515 (2006).

32. Gonzalez, G. A. Urban sprawl, global warming and the limits of ecological modernisation. *Environmental Politics* **14**, 344–362 (2005), doi:10.1080/0964410500087558.

33. Stern, S. M. Residential protectionism.

34. Harvey, D. The right to the city.

35. Stern, S. M. Residential protectionism, 1144.

36. Putnam, R. D. Bowling alone: America's declining social capital. *Journal of Democracy* **6**, 65–78 (1995).

37. Kelly, J. J. We shall not be moved: Urban communities, eminent domain and the socioeconomics of just compensation. *St. John's Legal Review* **80**, 960–961 (2006).

38. Stern, S. M. Residential protectionism.

39. DiPasquale, D. & Glaeser, E. L. Incentives and social capital: Are homeowners better citizens? *Journal of Urban Economics* **45**, 354–384 (1999), doi:10.1006/juec .1998.2098.

40. Stern, S. M. Residential protectionism.

41. Brown, D. L. Migration and community: Social networks in a multilevel world. *Rural Sociology* **67**, 1–23 (2002).

42. Gonzalez, G. A. Urban sprawl, global warming. Stern, S. M. Residential protectionism.

43. Jefferson, T. *Notes on the State of Virginia,* 164–165. W. W. Norton, 1787/1972.

44. Lands, L. Be a patriot, 949.

45. Lands, L. Be a patriot, 949.

46. Gonzalez, G. A. Urban sprawl, global warming.

47. Harvey, D. The right to the city.

48. Murie, A. Secure and contented citizens? Home ownership in Britain, in *Housing and Public Policy: Citizenship, Choice and Control* (ed. D. Mullins & A. Marsh), 79–98. Open University Press, 1998. Forrest, R., Murie, A. & Williams, P. *Home Ownership: Differentiation and Fragmentation.* Unwin Hyman, 1990.

49. Harvey, D. *A Brief History of Neoliberalism.* Oxford University Press, 2005.

50. Harvey, D. *A Brief History,* 159.

51. World Bank. *World Development Report 2009: Reshaping Economic Geography,* 206. World Bank/International Bank for Reconstruction and Development, 2009.

52. Harvey, D. The right to the city. Gonzalez, G. A. Urban sprawl, global warming. Harvey, D. *A Brief History.*

53. Stern, S. M. Residential protectionism.

54. Krugman, P. Home not-so-sweet home.

55. Krugman, P. Home not-so-sweet home.

56. Stern, S. M. Residential protectionism.

57. Smith, S. J., Easterlow, D., Munro, M. & Turner, K. M. Housing as health capital: How health trajectories and housing paths are linked. *Journal of Social Issues* **59**, 501–525 (2003).

58. Stern, S. M. Residential protectionism.

59. Harvey, D. The right to the city.

60. Harvey, D. The right to the city. Hanan, J. S. Home is where the capital is.

61. World Bank. *World Development Report 2009*, 206.

62. Myers, D. & Ryu, S. H. Aging baby boomers and the generational housing bubble: Foresight and mitigation of an epic transition. *Journal of the American Planning Association* **74**, 17–33 (2008).

63. Myers, D. & Ryu, S. H. Aging baby boomers.

64. Nelson, A. C. Leadership in a new era. *Journal of the American Planning Association* **72**, 393–407 (2006). Fishman, R. The fifth migration. *Journal of the American Planning Association* **71**, 357–367 (2005).

65. Nelson, A. C. Leadership in a new era.

66. Stern, S. M. Residential protectionism.

67. Harvey, D. *A Brief History*. Harvey, D. *MegaCities, Lecture 4*. Twynstra Gudde Management Consultants, 2000.

68. World Bank. *World Development Report 2009*.

69. Davis, M. *City of Quartz: Excavating the Future in Los Angeles*, 224. Verso, 1990.

70. Harvey, D. *MegaCities*.

71. Harvey, D. *MegaCities*.

Chapter 3 · "Housaholism" in the Greater Yellowstone Ecosystem

1. Ewing, S. My beautiful ranchette. *High Country News*, May 10 (1999), www.hcn.org/issues/154/4994/.

2. Gosnell, H., Haggerty, J. H. & Travis, W. R. Ranchland ownership change in the Greater Yellowstone Ecosystem, 1990–2001: Implications for conservation. *Society & Natural Resources* **19**, 743–758 (2006).

3. Noss, R. F., Carroll, C., Vance-Borland, K. & Wuerthner, G. A multicriteria assessment of the irreplaceability and vulnerability of sites in the Greater Yellowstone Ecosystem. *Conservation Biology* **16**, 895–908 (2002).

4. Driggs, B. W. *History of Teton Valley, Idaho*, 58. Eastern Idaho, 1970.

5. Minkin, T. America's coolest mountain towns. *Men's Journal* **August**, 48–56, at 54 (1996).

6. Jones, A. The 50 best places to live: Driggs, ID. *Men's Journal* **March**, 63–64, at 64 (2002).

7. The hot zones: Driggs, Idaho; Wilder side of the Tetons. *National Geographic Adventure* (2001), www.nationalgeographic.com/adventure/0107/trips_1.html.

8. Dworetzky, T. 83422: All eyes on Idaho's finest. *National Geographic* **February**, 128 (2003), http://ngm.nationalgeographic.com/ngm/0302/feature7/index.html.

9. Marin, R. Leave my town out of your "Top 10." *High Country News*, April 29 (2002), www.hcn.org/issues/225/11189/.

10. Sonoran Institute. *Growth Impacts in the Teton Region of Wyoming and Idaho.* Sonoran Institute, 2008.

11. Sonoran Institute. *Growth Impacts.*

12. Owen, D. *Green Metropolis: Why Living Smaller, Living Closer, and Driving Less Are the Keys to Sustainability.* Penguin, 2009.

13. Burger, B. M. & Carpenter, R. *Rural Real Estate Markets and Conservation Development in the Intermountain West: Perspectives, Challenges and Opportunities Emerging from the Great Recession.* Lincoln Institute of Land Policy, 2010.

14. Johnson, K. M. Unpredictable directions of rural population growth and migration, in *Challenges for Rural America in the Twenty-First Century* (ed. D. L. Brown & L. Swanson), 19–31. Pennsylvania State University Press, 2003. Shumway, J. M. & Davis, J. A. Nonmetropolitan population change in the Mountain West: 1910–1995. *Rural Sociology* **61**, 513–529 (1996). Smith, M. D. & Krannich, R. S. "Culture clash" revisited: Newcomer and longer-term residents' attitudes toward land use, development, and environmental issues in rural communities in the Rocky Mountain West. *Rural Sociology* **65**, 396–421 (2000).

15. Rothman, H. K. *Devil's Bargains: Tourism in the Twentieth-Century American West.* University Press of Kansas, 1998. Graber, E. E. Newcomers and oldtimers: Growth and change in a mountain town. *Rural Sociology* **39**, 504–513 (1974). Starrs, P. F. Conflict and change on the landscapes of the arid American West, in *The Changing American Countryside: Rural People and Places* (ed. E. M. Castle), 271–285. University Press of Kansas, 1995.

16. Starrs, P. F. Conflict and change.

17. Burger, B. M. & Carpenter, R. *Rural Real Estate Markets.*

18. Burger, B. M. & Carpenter, R. *Rural Real Estate Markets.*

19. Preston, G. *Fiscal Impacts of New Houses on Vacant Rural Subdivision Lots: Teton County, Idaho.* Lincoln Institute of Land Policy, Sonoran Institute, & RPI Consulting, 2010.

20. Burger, B. M. & Carpenter, R. *Rural Real Estate Markets.*

21. Preston, G. *Fiscal Impacts of New Houses.*

22. Preston, G. *Fiscal Impacts of New Houses.*

23. Noss, R. F., Carroll, C., Vance-Borland, K. & Wuerthner, G. A multicriteria assessment of the irreplaceability and vulnerability of sites in the Greater Yellowstone Ecosystem. *Conservation Biology* **16**, 895–908 (2002).

24. Noss, R. F. et al. A multicriteria assessment.

25. Idaho Fish & Game. Teton River trout population depressed, but improving. *Upper Snake Region Annual Fisheries Newsletter* **1**, 1–6 (2005).

26. Berger, J. The last mile: How to sustain long-distance migration in mammals. *Conservation Biology* **18**, 320–331 (2004).

27. Preston, G. *Fiscal Impacts of New Houses.*

28. Jones, A. The 50 best places to live, 64.

29. Dworetzky, T. 83422: All eyes, 128.

30. Smith, M. D. & Krannich, R. S. "Culture clash" revisited.

31. The study used a random sample of residents. Methodological details are described in Peterson, M. N., Chen, X. D. & Liu, J. G. Household location choices: Implications for biodiversity conservation. *Conservation Biology* 22, 912–921 (2008), doi:10.1111/j.1523-1739.2008.00929.x.

32. Peterson, M. N., Chen, X. D. & Liu, J. G. Household location choices.

33. Dunlap, R. E., Van Liere, K. D., Mertig, A. G. & Jones, R. E. Measuring endorsement of the new ecological paradigm: A revised NEP scale. *Journal of Social Issues* 56, 425–442 (2000).

34. Dunlap, R. E. et al. Measuring endorsement.

35. Kie, J. G. & Czech, B. Mule and black-tailed deer, in *Ecology and Management of Large Mammals in North America* (ed. S. Demarais & P. R. Krousman), 629–648. Prentice Hall, 2000. Skovlin, J. M. Habitat requirements and evaluation, in *Elk of North America: Ecology and Management* (ed. J. W. Thomas & D. E. Toweill), 369–413. Wildlife Management Institute, 1982.

36. Dietz, T., Stern, P. C. & Guagnano, G. A. Social structural and social psychological bases of environmental concern. *Environment and Behavior* 30, 450–471 (1998). Johnson, C. Y., Bowker, J. M. & Cordell, H. K. Ethnic variation in environmental belief and behavior: An examination of the new ecological paradigm in a social psychological context. *Environment and Behavior* 36, 157–186 (2004). Nord, M., Luloff, A. E. & Bridger, J. C. The association of forest recreation with environmentalism. *Environment and Behavior* 30, 235–246 (1998).

37. Hansen, A. J. et al. Effects of exurban development on biodiversity: Patterns, mechanisms, and research needs. *Ecological Applications* 15, 1893–1905 (2005).

38. On bird diversity, Marzluff, J. M. Island biogeography for an urbanizing world: How extinction and colonization may determine biological diversity in human-dominated landscapes. *Urban Ecosystems* 8, 157–177 (2005). Robinson, L., Newell, J. P. & Marzluff, J. A. Twenty-five years of sprawl in the Seattle region: Growth management responses and implications for conservation. *Landscape and Urban Planning* 71, 51–72 (2005). On fish species, Sass, G. G. et al. Fish community and food web responses to a whole-lake removal of coarse woody habitat. *Fisheries* 31, 321–330 (2006). On Key deer, Lopez, R. R. et al. Habitat use patterns of Florida Key deer: Implications of urban development. *Journal of Wildlife Management* 68, 900–908 (2004). On biodiversity conservation, Rasker, R. & Hansen, A. J. Natural amenities and population growth in the Greater Yellowstone region. *Human Ecology Review* 7, 30–40 (2000). On panda conservation, Liu, J. G. et al. A framework for evaluating the effects of human factors on wildlife habitat: The case of giant pandas. *Conservation Biology* 13, 1360–1370 (1999). Liu, J. G., Linderman, M., Ouyang, Z. & An, L. The pandas' habitat at Wolong Nature Reserve—response. *Science* 293, 603–605 (2001).

39. Liu, J. G. et al. Coupled human and natural systems. *Ambio* 36, 639–649 (2007).

40. Stroud, H. B. *The Promise of Paradise: Recreational and Retirement Communities in the United States since 1950.* Johns Hopkins University Press, 1995.

41. Lagro, J. A. Population growth beyond the urban fringe: Implications for rural land-use policy. *Landscape and Urban Planning* 28, 143–158 (1994). Shafer, C. L. A geography of hope: Pursuing the voluntary preservation of America's natural heritage. *Landscape and Urban Planning* 66, 127–171 (2004).

42. Mohai, P. Men, women, and the environment: An examination of the gender gap in environmental concern and activism. *Society and Natural Resources* **5**, 1–19 (1992). Milbrath, L. W. & Goel, M. L. *Political Participation: How and Why Do People Get Involved in Politics?* University Press of America, 1982.

43. Peterson, M. N. & Liu, J. G. Property rights and landscape planning in the Intermountain West: The Teton Valley case. *Landscape and Urban Planning* **86**, 126–133, (2008), doi:10.1016/j.landurbplan.2008.01.003.

44. Foucault, M. On the genealogy of ethics: An overview of work in progress, in *The Foucault Reader* (ed. P. Rabinow), 340–372. Pantheon, 1984.

45. Horwitz, M. J. *The Transformation of American Law, 1870–1960: The Crisis of Legal Orthodoxy*, 272. Oxford University Press, 1992.

46. Burger, B. M. & Carpenter, R. *Rural Real Estate Markets.*

47. Elliott, D. Premature subdivisions and what to do about them. Working paper, Lincoln Institute Product Code WP10DE1. Lincoln Institute of Land Policy, 2010. Preston, G. *Fiscal Impacts.*

48. Peterson, M. J. et al. Obscuring ecosystem function with application of the ecosystem services concept. *Conservation Biology* **24**, 113–119, (2010), doi:10.1111/j.1523-1739.2009.01305.x.

49. Elliott, D. Premature subdivisions.

50. Radeloff, V. C. et al. Housing growth in and near United States protected areas limits their conservation value. *Proceedings of the National Academy of Sciences of the United States of America* **107**, 940–945 (2010), doi:10.1073/pnas.0911131107.

51. Compas, E. Measuring exurban change in the American West: A case study in Gallatin County, Montana, 1973–2004. *Landscape and Urban Planning* **82**, 56–65, (2007), doi:10.1016/j.landurbplan.2007.01.016.

Chapter 4 · *Household Dynamics and Giant Panda Conservation*

1. Vitousek, P. M., Mooney, H. A., Lubchenco, J. & Melillo, J. M. Human domination of Earth's ecosystems. *Science* **277**, 494 (1997). United Nations. *The Millennium Development Goals Report, 2010.* United Nations, 2010.

2. McNeely, J. A. & Miller, K. R. IUCN, national parks, and protected areas: Priorities for action. *Environmental Conservation* **10**, 13–21 (1983). Armesto, J. J. Conservation targets in South American temperate forests. *Science* **282**, 1271–1272 (1998). Liu, J. et al. Ecological degradation in protected areas: The case of Wolong Nature Reserve for giant pandas. *Science* **292**, 98–101 (2001). Kramer, R. A., Schaik, C. V. & Johnson, J. *Last Stand: Protected Areas and the Defense of Tropical Biodiversity.* Oxford University Press, 1997. Dompka, V. *Human Population, Biodiversity and Protected Areas: Science and Policy Issues.* American Association for the Advancement of Science, 1996. Liu, J., Linderman, M., Ouyang, Z. & An, L. The pandas' habitat at Wolong Nature Reserve— a response. *Science* **293**, 603–604 (2001).

3. Zhang, H., Li, D., Wei, R., Tang, C. & Tu, J. Advances in conservation and studies on reproductivity of giant pandas in Wolong. *Sichuan Journal of Zoology* **16**, 31–33 (1997). State Forestry Administration. *The 3rd National Survey Report on Giant Panda in China.* Science, 2006.

4. Schaller, G. B., Hu, J., Pan, W. & Zhu, J. *The Giant Pandas of Wolong.* University of Chicago Press, 1985.

5. Tan, Y., Ouyang, Z. & Zhang, H. Spatial characteristics of biodiversity in Wolong Nature Reserve. *China's Biosphere Reserve* 3, 19–24 (1995).

6. He, N., Liang, C. & Yin, X. Sustainable community development in Wolong Nature Reserve. *Ecological Economy* 1, 15–23 (1996). Myers, N. et al. Biodiversity hotspots for conservation priorities. *Nature* 403, 853–858 (2000).

7. Liu, J. et al. Beyond population size: Examining intricate interactions among population structure, land use and environment in Wolong Nature Reserve (China), in *New Research on Population and Environment* (ed. B. Entwisle & P. Stern), 217–237. National Academy of Sciences Press, 2005. Schmidt, P. M. & M. J. Peterson. Biodiversity conservation and indigenous land management in the era of self-determination. *Conservation Biology* 23, 1458–1466 (2009).

8. Liu, J. et al. Beyond population size.

9. An, L., Lupi, F., Liu, J., Linderman, M. A. & Huang, J. Modeling the choice to switch from fuelwood to electricity: Implications for giant panda habitat conservation. *Ecological Economics* 42, 445–457 (2002).

10. Liu, J. et al. A framework for evaluating effects of human factors on wildlife habitats: The case on the giant pandas. *Conservation Biology* 13, 1360–1370 (1999).

11. Liu, J. et al. A framework for evaluating effects.

12. Bearer, S. et al. Effects of fuelwood collection and timber harvesting on giant panda habitat use. *Biological Conservation* 141, 385–393 (2008).

13. Liu, J. et al. Beyond population size.

14. An, L. & Liu, J. Long-term effects of family planning and other determinants of fertility on population and environment: Agent-based modeling evidence from Wolong Nature Reserve, China. *Population and Environment* 31, 427–459 (2010).

15. Liu, J. et al. A new paradigm for panda research and conservation: Integrating ecology with human demography, behavior, and socioeconomics, in *Giant Panda: Conservation Priorities for the 21st Century* (ed. D. G. Lindburg & K. Baragona), 217–225. University of California Press, 2004.

16. Liu, J. et al. A framework for evaluating effects.

17. Schaller, G. B. et al. *The Giant Pandas*. Liu, J. et al. A framework for evaluating effects.

18. Liu, J. et al. Human impacts on land cover and panda habitat in Wolong Nature Reserve: Linking ecological, socioeconomic, demographic, and behavioral data, in *People and the Environment: Approaches for Linking Household and Community Surveys to Remote Sensing and GIS* (ed. J. Fox, V. Mishra, R. Rindfuss & S. Walsh), 241–263. Kluwer Academic, 2003.

19. An, L. et al. Modeling the choice.

20. Liu, J. et al. A framework for evaluating effects. Liu, J. Integrating ecology with human demography, behavior, and socioeconomics: Needs and approaches. *Ecological Modelling* 140, 1–8 (2001).

21. He, G. et al. Spatial and temporal patterns of fuelwood collection in Wolong Nature Reserve: Implications for panda conservation. *Landscape and Urban Planning* 92, 1–9 (2009).

22. Liu, J., Daily, G. C., Ehrlich, P. R. & Luck, G. W. Effects of household dynamics on resource consumption and biodiversity. *Nature* 421, 530–533 (2003).

23. Liu, J., Ouyang, Z., Tan, Y., Yang, J. & Zhou, S. Changes in human population structure and implications for biodiversity conservation. *Population and Environment* **21**, 45–58 (1999). An, L., Mertig, A. G. & Liu, J. Adolescents leaving parental home: Psychosocial correlates and implications for conservation. *Population and Environment* **24**, 415–444 (2003).

24. Liu, J. et al. Effects of household dynamics.

25. An, L., Mertig, A. G. & Liu, J. Adolescents leaving parental home.

26. Liu, J. et al. Changes in human population structure.

27. An, L., Mertig, A. G. & Liu, J. Adolescents leaving parental home. Liu, J. et al. Ecological degradation in protected areas: The case of Wolong Nature Reserve for giant pandas. *Science* **292**, 98–101 (2001).

28. Goldscheider, F. K. & Goldscheider, C. *Leaving Home before Marriage: Ethnicity, Familism, and Generational Relationships.* University of Wisconsin Press, 1993. Goldscheider, F. K. & DaVanzo, J. Living arrangements and the transition to adulthood. *Demography* **22**, 545–563 (1985).

29. An, L., Mertig, A. G. & Liu, J. Adolescents leaving parental home.

30. Liu, J. et al. Beyond population size.

31. Liu, J. G. et al. Complexity of coupled human and natural systems. *Science* **317**, 1513–1516 (2007). Liu, J., Li, S., Ouyang, Z., Tam, C. & Chen, X. Ecological and socioeconomic effects of China's policies for ecosystem services. *Proceedings of the National Academy of Sciences of the United States of America* **105**, 9477–9482 (2008), doi:10.1073/pnas.0706436105.

32. Sichuan Forestry Survey Institute & Wolong Nature Reserve. *Forest Monitoring for Natural Forest Conservation Program in Wolong Nature Reserve.* Report, 2000.

33. Liu, J. et al. Beyond population size.

34. He, G. et al. Spatial and temporal patterns.

35. An, L. et al. Simulating demographic and socioeconomic processes on household level and implications for giant panda habitats. *Ecological Modelling* **140**, 31–50 (2001).

36. Liu, J. et al. Beyond population size.

37. Linderman, M. A. et al. Modeling the spatio-temporal dynamics and interactions of households, landscapes, and giant panda habitat. *Ecological Modelling* **183**, 47–65 (2005).

38. An, L. & Liu, J. Long-term effects of family planning.

39. An, L. et al. Simulating demographic and socioeconomic processes.

40. An, L. et al. Modeling the choice.

41. Liu, J. et al. Ecological and socioeconomic effects. Viña, A. et al. Temporal changes in giant panda habitat connectivity across boundaries of Wolong Nature Reserve, China. *Ecological Applications* **17**, 1019–1030 (2007).

42. Liu, J. et al. Protecting China's biodiversity. *Science* **300**, 1240–1241 (2003).

43. Yin, Y. P., Wang, F. W. & Sun, P. Landslide hazards triggered by the 2008 Wenchuan earthquake, Sichuan, China. *Landslides* **6**, 139–152 (2009).

44. Sichuan Department of Forestry. *Overall Planning for Post-Wenchuan Earthquake Restoration and Reconstruction in Wolong National Nature Reserve.* Sichuan Department of Forestry, 2008. Wenchuan County People's Government, Guangdong Province People's Government & Guangdong Province Institute of Planning and De-

sign. *Overall Planning for Post-Earthquake Restoration and Reconstruction in Wench-uan County.* Governmental report, 2008.

45. Wenchuan County People's Government, Guangdong Province People's Government & Guangdong Province Institute of Planning and Design. *Overall Planning.* State Planning Group of Post-Wenchuan Earthquake Restoration and Reconstruction. *The State Overall Planning for Post-Wenchuan Earthquake Restoration and Reconstruction.* Governmental report, 2008.

46. Hemin Zhang, Director of Wolong Nature Reserve, pers. obs.

47. Li, D. *Wolong Reconstruction Plan.* Beijing University, 2009.

48. Sichuan Department of Forestry. *Overall Planning.*

49. An, L. et al. Modeling the choice.

50. Liu, J. et al. The pandas' habitat.

51. Liu, J. et al. A framework for evaluating effects.

52. Liu, J. et al. Protecting China's biodiversity.

53. Agrawal, A. & Redford, K. Conservation and displacement: An overview. *Conservation and Society* 7 (2009).

Chapter 5 · Defusing the Housing Bomb with Your House

1. Gardner, G. T. & Stern, P. C. The short list: The most effective actions U.S. households can take to curb climate change. *Environment: Science and Policy for Sustainable Development* 50, 12–25 (2008).

2. Arnold, M. *Fuelwood Revisited: What Has Changed in the Last Decade?* Center for International Forestry Research, 2003.

3. Liu, J. G. et al. A framework for evaluating the effects of human factors on wildlife habitat: The case of giant pandas. *Conservation Biology* 13, 1360–1370 (1999).

4. Gardner, G. T. & Stern, P. C. The short list.

5. Van Haaren, R., Themelis, N. & Goldstein, N. The state of garbage in America. *BioCycle* 51, 16 (2010).

6. Van Haaren, R., Themelis, N. & Goldstein, N. The state of garbage.

7. U.S. Environmental Protection Agency. Wastes, in *U.S. Greenhouse Gas Emissions Inventory, 2010,* chapter 8. U.S. Environmental Protection Agency, 2010. Lou, X. & Nair, J. The impact of landfilling and composting on greenhouse gas emissions: A review. *Bioresource Technology* 100, 3792–3798 (2009).

8. U.S. Environmental Protection Agency. Wastes.

9. Van Haaren, R., Themelis, N. & Goldstein, N. The state of garbage.

10. Nowak, D. J. & Walton, J. T. Projected urban growth (2000–2050) and its estimated impact on the US forest resource. *Journal of Forestry* 103, 383–389 (2005).

11. Carter, T. & Fowler, L. Establishing green roof infrastructure through environmental policy instruments. *Environmental Management* 42, 151–164 (2008). Alberti, M. *Advances in Urban Ecology: Integrating Humans and Ecological Processes in Urban Ecosystems.* Springer Verlag, 2008.

12. Nassauer, J. I., Wang, Z. & Dayrell, E. What will the neighbors think? Cultural norms and ecological design. *Landscape and Urban Planning* 92, 282–292 (2009).

13. Czech, B., Krausman, P. R. & Devers, P. K. Economic associations among causes of species endangerment in the United States. *Bioscience* 50, 593–601 (2000).

14. Allcott, H. & Mullainathan, S. Behavior and energy policy. *Science* **327**, 1204 (2010).

15. Tranter, P. *Effective Speeds: Car Costs Are Slowing Us Down*. Australian Greenhouse Office, 2004, www.environment.gov.au/archive/settlements/transport/publica tions/effectivespeeds.html, accessed January 11, 2013.

16. Granade, H. C. et al. *Unlocking Energy Efficiency in the U.S. Economy*. McKinsey, 2009.

17. Gardner, G. T. & Stern, P. C. The short list.

18. Granade, H. C. et al. *Unlocking Energy Efficiency*.

19. Granade, H. C. et al. *Unlocking Energy Efficiency*.

20. Davis, S. C., Diegel, S. W. & Boundy, E. G. *Transportation Energy Data Book: Edition 30*. Oak Ridge National Laboratory, 2011.

21. American Public Transportation Association. *Public Transportation Factbook*. American Public Transportation Association, 2008. Davis, S. C., Diegel, S. W. & Boundy, E. G. *Transportation Energy Data Book*.

22. Davis, S. C., Diegel, S. W. & Boundy, E. G. *Transportation Energy Data Book*.

23. Goodman, J. D. An electric boost for bicyclists. *New York Times*, February 1 (2010), B1, www.nytimes.com/2010/02/01/business/global/01ebike.html?_r=2/.

24. Davis, S. C., Diegel, S. W. & Boundy, E. G. *Transportation Energy Data Book*.

25. Tranter, P. *Effective Speeds*.

26. Andersen, L. B., Schnohr, P., Schroll, M. & Hein, H. O. All-cause mortality associated with physical activity during leisure time, work, sports, and cycling to work. *Archives of Internal Medicine* **160**, 1621 (2000).

27. U.S. Department of Transportation, National Highway Traffic Safety Administration. *Traffic Safety Facts (2009 Data): Bicyclists and Other Cyclists*. NHTSA National Center for Statistics & Analysis, 2009.

28. Jacobsen, P. L. Safety in numbers: More walkers and bicyclists, safer walking and bicycling. *Injury Prevention* **9**, 205–209 (2003).

29. Andersen, L. B. et al. All-cause mortality.

30. U.S. Department of Transportation. *Traffic Safety Facts (2009 Data)*.

31. U.S. Environmental Protection Agency, Office of Water. *Water On Tap: What You Need to Know*. U.S. Environmental Protection Agency, 2009.

32. U.S. Environmental Protection Agency. *Water Efficiency Technology Fact Sheet: Composting Toilets*. U.S. Environmental Protection Agency, 1999.

33. Owen, D. *Green Metropolis: Why Living Smaller, Living Closer, and Driving Less Are Keys to Sustainability*. Penguin, 2009.

34. Breuste, J. H. Decision making, planning and design for the conservation of indigenous vegetation within urban development. *Landscape and Urban Planning* **68**, 439–452 (2004). Grimm, N. B. et al. Global change and the ecology of cities. *Science* **319**, 756 (2008).

35. Zhou, W., Troy, A., Grove, J. M. & Jenkins, J. C. Can money buy green? Demographic and socioeconomic predictors of lawn-care expenditures and lawn greenness in urban residential areas. *Society and Natural Resources* **22**, 744–760 (2009).

36. Robbins, P. & Birkenholtz, T. Turfgrass revolution: Measuring the expansion of the American lawn. *Land Use Policy* **20**, 181–194 (2003). Milesi, C. et al. Mapping

and modeling the biogeochemical cycling of turfgrasses in the United States. *Environmental Management* **36**, 426–438 (2005).

37. Robbins, P. & Sharp, J. T. Producing and consuming chemicals: The moral economy of the American lawn, in *Urban Ecology: An International Perspective on the Interactions between Humans and Nature* (ed. J. M. Marzluff et al.), 180–206. Springer, 2008.

38. Kaye, J. P., Groffman, P. M., Grimm, N. B., Baker, L. A. & Pouyat, R. V. A distinct urban biogeochemistry? *Trends in Ecology & Evolution* **21**, 192–199 (2006).

39. Zhou, W. et al. Can money buy green? Bijoor, N. S., Czimczik, C. I., Pataki, D. E. & Billings, S. A. Effects of temperature and fertilization on nitrogen cycling and community composition of an urban lawn. *Global Change Biology* **14**, 2119–2131 (2008).

40. Adams, C. E. & Lindsey, K. J. *Urban Wildlife Management.* CRC Press, 2010.

41. Helfand, G. E., Sik Park, J., Nassauer, J. I. & Kosek, S. The economics of native plants in residential landscape designs. *Landscape and Urban Planning* **78**, 229–240 (2006). Martin, C. A., Peterson, K. A. & Stabler, L. B. Residential landscaping in Phoenix, Arizona, U.S.: Practices and preferences relative to covenants, codes, and restrictions. *Journal of Arboriculture* **29**, 9–17 (2003). Troy, A. R., Grove, J. M., O'Neil-Dunne, J. P. M., Pickett, S. T. A. & Cadenasso, M. L. Predicting opportunities for greening and patterns of vegetation on private urban lands. *Environmental Management* **40**, 394–412 (2007). Grove, J. et al. Characterization of households and its implications for the vegetation of urban ecosystems. *Ecosystems* **9**, 578–597 (2006).

42. Nassauer, J. I., Wang, Z. & Dayrell, E. What will the neighbors think?

43. Allen, S. C., Moorman, C. E., Peterson, M. N., Hess, G. R. & Moore, S. E. Overcoming socio-economic barriers to conservation subdivisions: A case-study of four successful communities. *Landscape and Urban Planning* **106**, 244–252 (2012).

44. Jorgensen, A., Hitchmough, J. & Dunnett, N. Woodland as a setting for housing-appreciation and fear and the contribution to residential satisfaction and place identity in Warrington New Town, UK. *Landscape and Urban Planning* **79**, 273–287 (2007), doi:10.1016/j.landurbplan.2006.02.015. Kaplan, R. & Austin, M. E. Out in the country: Sprawl and the quest for nature nearby. *Landscape and Urban Planning* **69**, 235–243 (2004). Todorova, A., Asakawa, S. & Aikoh, T. Preferences for and attitudes toward street flowers and trees in Sapporo, Japan. *Landscape and Urban Planning* **69**, 403–416 (2004), doi:10.1016/j.landurbplan.2003.11.001.

45. On birds, Moorman, C., Johns, M., Thomas Bowen, L. & Gerwin, J. *Managing Backyards and Other Urban Habitats for Birds.* Publication AG-636-1. North Carolina Cooperative Extension Service, 2002. On reptiles and amphibians, Anderson, J., Beane, J. C., Hall, J. G. & Moorman, C. *Reptiles and Amphibians in Your Backyard.* Publication AG-744. North Carolina Cooperative Extension Service, 2011. On other wildlife, Moorman, C. E., Johns, M. & Thomas Bowen, L. *Landscaping for Wildlife with Native Plants.* Publication AG-636-3. North Carolina Cooperative Extension Service, 2002. Tallamy, D. W. & Darke, R. *Bringing Nature Home: How You Can Sustain Wildlife with Native Plants.* Timber Press, 2009.

46. Koh, L. P. & Sodhi, N. S. Importance of reserves, fragments, and parks for butterfly conservation in a tropical urban landscape. *Ecological Applications* **14**, 1695–1708 (2004).

47. Lepczyk, C. A., Mertig, A. G. & Liu, J. Assessing landowner activities related to birds across rural-to-urban landscapes. *Environmental Management* **33**, 110–125 (2004).

48. Schmidt, P. M., Lopez, R. R. & Collier, B. A. Survival, fecundity, and movements of free-roaming cats. *Journal of Wildlife Management* **71**, 915–919 (2007). Dauphiné, N. & Cooper, R. J. Impacts of free-ranging domestic cats (*Felis catus*) on birds in the United States: A review of recent research with conservation and management recommendations, in *Tundra to Tropics: Connecting Birds, Habitats and People; Proceedings of the Fourth International Partners in Flight Conference*, 205–219. Partners in Flight, 2009. O'Brien, S. & Johnson, W. The evolution of cats. *Scientific American* **297**, 68–75 (2007).

49. Audubon Society of Portland. *Cats and Wildlife*. Audubon Society of Portland, 2007, http://audubonportland.org/backyardwildlife/brochures/cats/catslanding/. Georgia Ornithological Society. *Georgia Ornithological Society Position Statement: Managing Feral and Free-Ranging Domestic Cats*. Georgia Ornithological Society, 2011, www.gos.org/orginfo/gos-cats.htm. Lepczyk, C. A. et al. What conservation biologists can do to counter trap-neuter-return: Response to Longcore et al. *Conservation Biology* **24**, 627–629 (2010).

50. Johns, R. & Bies, L. House cats as predators. *Wyoming Wildlife* **76**, 45 (2012).

51. Hess, S. C. By land and by sea—the widespread threat of feral cats on Hawaiian wildlife. *Wildlife Professional* **5**, 66–67 (2011).

52. Granade, H. C. et al. *Unlocking Energy Efficiency*.

53. Owen, D. *Green Metropolis*.

54. Gardner, G. T. & Stern, P. C. The short list.

55. Granade, H. C. et al. *Unlocking Energy Efficiency*.

56. Granade, H. C. et al. *Unlocking Energy Efficiency*.

57. Granade, H. C. et al. *Unlocking Energy Efficiency*.

58. Allcott, H. & Mullainathan, S. Behavior and energy policy

59. Allcott, H. & Mullainathan, S. Behavior and energy policy.

60. Tranter, P. *Effective Speeds*. Burrington, S. H. & Conservation Law Foundation. *Road Kill: How Solo Driving Runs Down the Economy*. Conservation Law Foundation, 1994.

61. Granade, H. C. et al. *Unlocking Energy Efficiency*.

62. Gardner, G. T. & Stern, P. C. The short list.

63. In some cases the "we" used in this section refers to Nils Peterson. Specifically, those instances relate to who was paying for the recommended changes needed to improve energy efficiency, benefiting from the changes, riding bikes, and crawling around in the attic with caulk and spray foam.

64. National Association of Home Builders. *Housing Facts, Figures and Trends 2004*. NAHB Advocacy / Public Affairs, in cooperation with the NAHB Economics Group, 2004.

65. Davis, S. C., Diegel, S. W. & Boundy, E. G. *Transportation Energy Data Book*.

66. People who use public transit walk more than people who commute by personal vehicles.

Chapter 6 · Individual and Local Strategies for Defusing the Housing Bomb

1. Rusk, D. Social framework: Sprawl, race, and concentrated poverty—changing the "rules of the game," in *Urban Sprawl: A Comprehensive Reference Guide* (ed. D. C. Soule), 90–102. Greenwood Press, 2006.

2. On segregation, Rusk, D. Social framework. On obesity, Ewing, R., Schmid, T., Killingsworth, R., Zlot, A. & Raudenbush, S. Relationship between urban sprawl and physical activity, obesity, and morbidity, in *Urban Ecology: An International Perspective on the Interactions between Humans and Nature* (ed. J. M. Marzluff et al.), 567–582. Springer, 2008. Papas, M. A. et al. The built environment and obesity. *Epidemiologic Reviews* 29, 129 (2007). On social capital, Putnam, R. D. Bowling alone: America's declining social capital. *Journal of Democracy* 6, 65–78 (1995). On wildlife species, Czech, B., Krausman, P. R. & Devers, P. K. Economic associations among causes of species endangerment in the United States. *Bioscience* 50, 593–601 (2000). On oil supplies, Hirsch, R., Bezdek, R. & Wendling, R. *Peaking of World Oil Production: Impacts, Mitigation, & Risk Management*. Report presented to the U.S. Department of Energy, 2005.

3. McGinn, D. F. *House Lust: America's Obsession with Our Homes*. Random House, 2008.

4. McGinn, D. F. *House Lust*.

5. Ewing, R., Pendall, R. & Chen, D. *Measuring Sprawl and Its Impact*. Smart Growth America, 2002, www.smartgrowthamerica.org/research/measuring-sprawl -and-its-impact/, accessed January 3, 2013.

6. Peterson, M. N., Thurmond, B., Mchale, M., Rodriguez, S., Bondell, H.D. & Cook, M. 2012. Predicting native plant landscaping preferences in urban areas. *Sustainable Cities and Society* 5, 70–76.

7. Cialdini, R. B., Kallgren, C. A. & Reno, R. R. A focus theory of normative conduct: A theoretical refinement and reevaluation of the role of norms in human behavior. *Advances in Experimental Social Psychology* 24, 201–234 (1991).

8. Cialdini, R. B. & Trost, M. R. Social influence: Social norms, conformity, and compliance, in *The Handbook of Social Psychology*, vol. 2, 4th ed. (ed. D. T. Gilbert, S. T. Fiske & G. Lindzey), 151–192. McGraw-Hill, 1998. Fishbein, M. & Ajzen, I. *Belief, Attitude, Intention, and Behavior: An Introduction to Theory and Research*. Addison-Wesley, 1975.

9. Leinberger, C. B. & Alfonzo, M. *Walk This Way: The Economic Promise of Walkable Places in Metropolitan Washington, D.C.* Brookings Institution, 2012.

10. Flint, A. *This Land: The Battle over Sprawl and the Future of America*. Johns Hopkins University Press, 2006.

11. Warner, J. T. & Pleeter, S. The personal discount rate: Evidence from military downsizing programs. *American Economic Review*, 33–53 (2001). Meier, A. K. & Whittier, J. Consumer discount rates implied by purchases of energy-efficient refrigerators. *Energy* 8, 957–962 (1983).

12. Boyce, C. J., Brown, G. D. A. & Moore, S. C. Money and happiness. *Psychological Science* 21, 471 (2010).

13. Mason, R. Conspicuous consumption and the positional economy: Policy and prescription since 1970. *Managerial and Decision Economics* 21, 123–132 (2000).

14. Allcott, H. & Mullainathan, S. Behavior and energy policy. *Science* 327, 1204 (2010).

15. Lang, R. & LeFurgy, J. B. *Boomburbs: The Rise of America's Accidental Cities*. Brookings Institution, 2007.

16. Beck, U. *Risk Society: Towards a New Modernity*, trans. M. Ritter. Sage, 1992.

17. Kunstler, J. H. *The Geography of Nowhere: The Rise and Decline of America's Man-Made Landscape.* Free Press, 1994. Kunstler, J. H. *The Long Emergency: Surviving the End of Oil, Climate Change, and Other Converging Catastrophes of the Twenty-First Century.* Grove Press, 2006.

18. Olshansky, S. J. et al. A potential decline in life expectancy in the United States in the 21st century. *New England Journal of Medicine* 352, 1138–1145 (2005). Ewing, R. et al. Relationship between urban sprawl and physical activity. Papas, M. A. et al. The built environment and obesity. *Epidemiologic Reviews* 29, 129 (2007).

19. Brown, L. R. *Plan B 3.0: Mobilizing to Save Civilization.* W. W. Norton, 2008.

20. Lang, R. & LeFurgy, J. B. *Boomburbs.*

21. Lang, R. & LeFurgy, J. B. *Boomburbs.*

22. Lang, R. & LeFurgy, J. B. *Boomburbs.*

23. Florida, R. Cities and the creative class. *City & Community* 2, 3–19 (2003).

24. Wickersham, J. Legal framework: The laws of sprawl and the laws of smart growth, in *Urban Sprawl* (ed. D. C. Soule), 26–60.

25. Jacobs, J. *The Death and Life of Great American Cities.* Vintage, 1961.

26. Duany, A., Speck, J. & Lydon, M. *The Smart Growth Manual.* McGraw-Hill, 2010. Emerine, D., Shenot, C., Bailey, M. K., Sobel, L. & Susman, M. *This Is Smart Growth.* International City/County Management Association & U.S. Environmental Protection Agency, 2006.

27. Dittmar, H. & Ohland, G. *The New Transit Town: Best Practices in Transit-Oriented Development.* Island Press, 2004. Calthorpe, P. *The Next American Metropolis: Ecology, Community, and the American Dream.* Princeton Architectural Press, 1993.

28. Tachieva, G. *Sprawl Repair Manual.* Island Press, 2010.

29. Mason, S. G. Can community design build trust? A comparative study of design factors in Boise, Idaho neighborhoods. *Cities* 27, 456–465 (2010).

30. Trudeau, D. & Malloy, P. Suburbs in disguise? Examining the geographies of the new urbanism. *Urban Geography* 32, 424–447 (2011).

31. Allen, S. C., Moorman, C. E., Peterson, M. N., Hess, G. R. & Moore, S. E. Overcoming socio-economic barriers to conservation subdivisions: A case-study of four successful communities. *Landscape and Urban Planning* 106, 244–252 (2012).

32. Arendt, R. & Harper, H. *Conservation Design for Subdivisions: A Practical Guide to Creating Open Space Networks.* Island Press, 1996.

33. Allen, S. C. et al. Overcoming socio-economic barriers.

34. Mohamed, R. The economics of conservation subdivisions' price premiums, improvement costs, and absorption rates. *Urban Affairs Review* 41, 376–399 (2006). Milder, J. C. A framework for understanding conservation development and its ecological implications. *Bioscience* 57, 757–768 (2007).

35. Mohamed, R. The economics of conservation subdivisions. Bowman, T. & Thompson, J. Barriers to implementation of low-impact and conservation subdivision design: Developer perceptions and resident demand. *Landscape and Urban Planning* 92, 96–105 (2009). Pejchar, L., Morgan, P. M., Caldwell, M. R., Palmer, C. & Daily, G. C. Evaluating the potential for conservation development: Biophysical, economic, and institutional perspectives. *Conservation Biology* 21, 69–78 (2007).

36. Grey-Ross, R., Downs, C. T. & Kirkman, K. Using housing estates as conser-

vation tools: A case study in KwaZulu-Natal, South Africa. *Applied Geography* **29**, 371–376 (2009).

37. Allen, S. C. et al. Overcoming socio-economic barriers.

38. Carter, T. Developing conservation subdivisions: Ecological constraints, regulatory barriers, and market incentives. *Landscape and Urban Planning* **92**, 117–124 (2009).

39. Allen, S. C. et al. Overcoming socio-economic barriers.

40. Maruyama, Y., Nishikido, M. & Iida, T. The rise of community wind power in Japan: Enhanced acceptance through social innovation. *Energy Policy* **35**, 2761–2769 (2007).

41. Wickersham, J. Legal framework. Allen, S. C. et l. Overcoming socio-economic barriers.

42. Ohm, B. W. & Sitkowski, R. J. Influence of new urbanism on local ordinances: The twilight of zoning? *Urban Lawyer* **35**, 783 (2003).

43. Allen, S. C. et al. Overcoming socio-economic barriers.

44. Wickersham, J. Legal framework.

45. Ben-Zadok, E. Consistency, concurrency and compact development: Three faces of growth management implementation in Florida. *Urban Studies* **42**, 2167 (2005).

46. Lerman, B. R. Mandatory inclusionary zoning: The answer to the affordable housing problem. *Boston College Environmental Affairs Law Review* **33**, 383 (2006).

47. Mukhija, V., Regus, L., Slovin, S. & Das, A. Can inclusionary zoning be an effective and efficient housing policy? Evidence from Los Angeles and Orange Counties. *Journal of Urban Affairs* **32**, 229–252 (2010). Ellickson, R. C. The irony of inclusionary zoning. *Southern California Law Review* **54**, 1167 (1980).

48. Mukhija, V. et al. Can inclusionary zoning?

49. Lerman, B. R. Mandatory inclusionary zoning.

50. Pucher, J., Dill, J. & Handy, S. Infrastructure, programs, and policies to increase bicycling: An international review. *Preventive Medicine* **50**, S106–S125 (2010).

51. Yim, S. H. L. & Barrett, S. R. H. Public health impacts of combustion emissions in the United Kingdom. *Environmental Science & Technology* **46**, 4291–4296 (2012).

52. Brown, L. R. *Plan B 3.0.*

53. Vanderkooy, Z. *Seville, Spain's Remarkable Transformation.* Green Lane Project, 2012, http://greenlaneproject.org/articles/view/76/.

54. Flusche, D. *The Economic Benefits of Bicycle Infrastructure Investments.* League of American Bicyclists, 2009.

55. Jun, M. J. Are Portland's smart growth policies related to reduced automobile dependence? *Journal of Planning Education and Research* **28**, 100–107 (2008).

56. Jun, M. J. Portland's smart growth policies.

57. Flusche, D. *The Economic Benefits.*

58. Yim, S. H. L. & Barrett, S. R. H. Public health impacts of combustion emissions in the United Kingdom. *Environmental Science & Technology* **46**, 4291–4296 (2012). Chen, H., Rufolo, A. & Dueker, K. J. Measuring the impact of light rail systems on single family home values: A hedonic approach with GIS application. Discussion paper 97-3. Portland State University, 1997.

59. Pucher, J., Dill, J. & Handy, S. Infrastructure, programs, and policies.

60. Pucher, J. & Buehler, R. Making cycling irresistible: Lessons from the Netherlands, Denmark and Germany. *Transport Reviews* **28**, 495–528 (2008).

61. Rosenbloom, S. Mobility of the elderly, in *Transportation in an Aging Society: A Decade of Experience; Technical Papers and Reports from a Conference, November 7–9, 1999, Bethesda, Maryland*, 3–21. Transportation Research Board, 2004, http://onlinepubs.trb.org/onlinepubs/conf/reports/cp_27.pdf.

62. Cobb, R. & Coughlin, J. Transportation policy for an aging society: Keeping older Americans on the move, in *Transportation in an Aging Society*, 275–289.

63. Rosenbloom, S. Mobility of the elderly.

64. Flusche, D. *The Economic Benefits*.

65. Podobnik, B. New urbanism and the generation of social capital: Evidence from Orenco Station. *National Civic Review* **91**, 245–255 (2002).

66. Yang, Y. A tale of two cities: Physical form and neighborhood satisfaction in metropolitan Portland and Charlotte. *Journal of the American Planning Association* **74**, 307–323 (2008).

67. Kaplowitz, M. D., Machemer, P. & Pruetz, R. Planners' experiences in managing growth using transferable development rights (TDR) in the United States. *Land Use Policy* **25**, 378–387 (2008).

68. Fischel, W. A. *The Homevoter Hypothesis: How Home Values Influence Local Government Taxation, School Finance, and Land-Use Policies*. Harvard University Press, 2005.

69. Fischel, W. A. Homevoters, municipal corporate governance, and the benefit view of the property tax. *National Tax Journal* **54**, 157–174 (2001).

70. McNamara, M. A. Legality and efficacy of homeowner's equity assurance: A study of Oak Park, Illinois. *Northwestern University Law Review* **78**, 1463 (1983).

71. Goetzmann, W. et al. Home equity insurance: A pilot project. Yale ICF working paper 03-12, 2003, http://papers.ssrn.com/sol3/papers.cfm?abstract_id=410141/.

72. McNamara, M. A. Legality and efficacy. Goetzmann, W. et al. Home equity insurance.

73. McNamara, M. A. Legality and efficacy. Goetzmann, W. et al. Home equity insurance.

74. Krizek, K. J. et al. *Guidelines for Analysis of Investments in Bicycle Facilities*. Transportation Research Board, 2006.

75. Lang, R. & LeFurgy, J. B. *Boomburbs*.

76. Soule, D. C. (ed.). *Urban Sprawl*.

77. Trudeau, D. & Malloy, P. Suburbs in disguise? Examining the geographies of the new urbanism. *Urban Geography* **32**, 424–447 (2011).

78. Allen, S. C. et al. Overcoming socio-economic barriers.

79. Endres, D., Sprain, L. M. & Peterson, T. R. *Social Movement to Address Climate Change: Local Steps for Global Action*. Cambria Press, 2009.

Chapter 7 · Large-Scale Strategies for Defusing the Housing Bomb

1. Wickersham, J. Legal framework: The laws of sprawl and the laws of smart growth, in *Urban Sprawl: A Comprehensive Reference Guide* (ed. D. C. Soule), 26–60. Greenwood Press, 2006.

2. Chad Anderson, a biologist at the Key Deer National Wildlife Refuge, provided this insight.

3. Martin, S. L., Lee, S. M. & Lowry, R. National prevalence and correlates of walking and bicycling to school. *American Journal of Preventive Medicine* **33**, 98–105 (2007).

4. Andersen, L. B., Schnohr, P., Schroll, M. & Hein, H. O. All-cause mortality associated with physical activity during leisure time, work, sports, and cycling to work. *Archives of Internal Medicine* **160**, 1621 (2000).

5. Weitz, J. From quiet revolution to smart growth: State growth management programs, 1960 to 1999. *Journal of Planning Literature* **14**, 266–337 (1999).

6. Fischlein, M. et al. Policy stakeholders and deployment of wind power in the sub-national context: A comparison of four U.S. states. *Energy Policy* **38**, 4429–4439 (2010).

7. Wickersham, J. H. Quiet revolution continues: The emerging new model for state growth management statutes. *Harvard Environmental Law Review* **18**, 489 (1994).

8. Wickersham, J. Legal framework.

9. Wickersham, J. Legal framework.

10. Jun, M. J. The effects of Portland's urban growth boundary on urban development patterns and commuting. *Urban Studies* **41**, 1333–1348 (2004).

11. Rudel, T. K., O'Neill, K., Gottlieb, P., McDermott, M. & Hatfield, C. From middle to upper class sprawl? Land use controls and changing patterns of real estate development in northern New Jersey. *Annals of the Association of American Geographers* **101**, 609–624 (2011).

12. Rudel, T. K. et al. From middle to upper class sprawl?.

13. Gonzalez, G. A. Urban sprawl, global warming and the limits of ecological modernisation. *Environmental Politics* **14**, 344–362 (2005). Gonzalez, G. A. *Urban Sprawl, Global Warming, and the Empire of Capital.* State University of New York Press, 2009.

14. McLaughlin, N. A. Increasing the tax incentives for Conservation Easement Donations: A responsible approach. *Ecology Law Quarterly* **31**, 1–115 (2004).

15. McLaughlin, N. A. Increasing the tax incentives.

16. These figures are based on adjusted gross incomes of $250,000, $75,000, and $35,000 a year for high, medium, and low incomes, respectively. To maximize benefits, the total tax savings were spread over six years.

17. Environmental Defense Fund. *Smart Grid: Revolutionizing Our Energy Future.* Environmental Defense Fund, 2012, www.edf.org/energy/smart-grid-overview/. McDaniel, P. & McLaughlin, S. Security and privacy challenges in the smart grid. *Security & Privacy, IEEE* **7**, 75–77 (2009).

18. Katzenstein, W., Fertig, E. & Apt, J. The variability of interconnected wind plants. *Energy Policy* **38**, 4400–4410 (2010).

19. Peterson, S. B., Whitacre, J. & Apt, J. The economics of using plug-in hybrid electric vehicle battery packs for grid storage. *Journal of Power Sources* **195**, 2377–2384 (2010).

20. Apt, J. & Fischhoff, B. Power and people. *Electricity Journal* **19**, 17–25 (2006), doi:10.1016/j.tej.2006.09.008.

21. Karathodorou, N., Graham, D. J. & Noland, R. B. Estimating the effect of urban density on fuel demand. *Energy Economics* **32**, 86–92 (2010).

22. Karathodorou, N., Graham, D. J. & Noland, R. B. Estimating the effect of urban density.

23. Sassi, F. *Obesity and the Economics of Prevention: Fit not Fat.* Organisation for Economic Co-operation and Development, 2010.

24. Wang, Y., Beydoun, M. A., Liang, L., Caballero, B. & Kumanyika, S. K. Will all Americans become overweight or obese? Estimating the progression and cost of the US obesity epidemic. *Obesity* **16**, 2323–2330 (2008).

25. Bemelmans-Videc, M. L., Rist, R. C. & Vedung, E. *Carrots, Sticks, and Sermons: Policy Instruments and Their Evaluation.* Transaction, 2003.

26. Poel, B., van Cruchten, G. & Balaras, C. A. Energy performance assessment of existing dwellings. *Energy and Buildings* **39**, 393–403 (2007).

27. Flusche, D. *The Economic Benefits of Bicycle Infrastructure Investments.* League of American Bicyclists, 2009.

28. Flusche, D. *The Economic Benefits.*

29. Flusche, D. *The Economic Benefits.*

30. Federal Highway Administration. *Report to the U.S. Congress on the Outcomes of the Nonmotorized Transportation Pilot Program SAFETEA-LU, Section 1807.* Federal Highway Administration, 2012.

31. Davis, S. C., Diegel, S. W. & Boundy, E. G. *Transportation Energy Data Book: Edition 30.* Oak Ridge National Laboratory, 2011.

32. Buehler, R., Pucher, J., Merom, D. & Bauman, A. Active travel in Germany and the U.S.: Contributions of daily walking and cycling to physical activity. *American Journal of Preventive Medicine* **41**, 241–250 (2011).

33. Buehler, R. et al. Active travel in Germany.

34. Meyers, S., Williams, A. & Chan, P. *Energy and Economic Impacts of U.S. Federal Energy and Water Conservation Standards Adopted from 1987 through 2010.* Lawrence Berkeley National Laboratory, 2011.

35. Meyers, S., Williams, A. & Chan, P. *Energy and Economic Impacts.*

36. Meyers, S., Williams, A. & Chan, P. *Energy and Economic Impacts.*

37. Committee on the Effectiveness and Impact of Corporate Average Fuel Economy Standards. *Effectiveness and Impact of Corporate Average Fuel Economy (CAFE) Standards.* National Academies Press, 2002.

38. Weiss, J., Dunkelberg, E. & Vogelpohl, T. Improving policy instruments to better tap into homeowner refurbishment potential: Lessons learned from a case study in Germany. *Energy Policy* **44**, 1–490 (2012).

39. Mutter, M. et al. *The Slum Upgrading Facility Handbook.* United Nations, Human Settlements Programme, 2006. Farha, L. *Forced Evictions: Global Crisis, Global Solutions.* United Nations, 2011.

40. United Nations. *The Millennium Development Goals Report, 2011.* United Nations, 2011.

41. Mutter, M. et al. *The Slum Upgrading Facility Handbook.*

42. Giroux, H. *The Terror of Neoliberalism: Authoritarianism and the Eclipse of Democracy.* Paradigm, 2004. Harvey, D. *A Brief History of Neoliberalism.* Oxford University Press, 2005.

43. Mutter, M. et al. *The Slum Upgrading Facility Handbook,* 5.

44. Du Plessis, J. The growing problem of forced evictions and the crucial impor-

tance of community-based, locally appropriate alternatives. *Environment and Urbanization* **17**, 123–134 (2005).

45. Du Plessis, J. The growing problem of forced evictions. Farha, L. *Forced Evictions.*

46. Harvey, D. The right to the city. *New Left Review* **53**, 23–40 (2008).

47. Harvey, D. *A Brief History of Neoliberalism.*

48. Du Plessis, J. The growing problem of forced evictions. Farha, L. *Forced Evictions.*

49. Du Plessis, J. The growing problem of forced evictions. Farha, L. *Forced Evictions.*

50. Farha, L. *Forced Evictions.* Agrawal, A. & Redford, K. Conservation and displacement: An overview. *Conservation and Society* **7**, 1 (2009). Harvey, D. The right to the city.

51. Rydin, Y. et al. Shaping cities for health: Complexity and the planning of urban environments in the 21st century. *Lancet* **379**, 2079–2108 (2012).

52. Rydin, Y. et al. Shaping cities for health.

53. Harvey, D. *A Brief History of Neoliberalism.* Harvey, D. *MegaCities, Lecture 4.* Twynstra Gudde Management Consultants, 2000.

54. Rydin, Y. et al. Shaping cities for health.

55. Rydin, Y. et al. Shaping cities for health.

56. World Health Organization. *Health in the Green Economy: Co-Benefits to Health of Climate Change Mitigation.* World Health Organization, 2011.

57. Rydin, Y. et al. Shaping cities for health.

58. O'Neill, R. *Global Smart Grid Federation Report, 2012.* Global Smart Grid Federation, 2010, www.globalsmartgridfederation.org.

59. Peterson, M. J., Hall, D. M., Feldpausch-Parker, A. M. & Peterson, T. R. Obscuring ecosystem function with application of the ecosystem services concept. *Conservation Biology* **24**, 113–119 (2010), doi:10.1111/j.1523-1739.2009.01305.x.

Conclusion

1. Mertig, A. G., Dunlap, R. E. & Morrison, D. E. The environmental movement in the United States, in *Handbook of Environmental Sociology* (ed. R. E. Dunlap & W. Michelson), 448–481. Greenwood Press, 2002.

2. Diamond, J. M. *Collapse: How Societies Choose to Fail or Succeed.* Viking, 2005. Liu, J. G. & Diamond, J. China's environment in a globalizing world. *Nature* **435**, 1179–1186 (2005). Liu, J. G. et al. Coupled human and natural systems. *Ambio* **36**, 639–649 (2007). Liu, J. et al. Framing sustainability in a telecoupled world. *Ecology and Society* **18**, 26 (2013). www.ecologyandsociety.org/vol18/iss2/art26/.

3. Czech, B., Krausman, P. R. & Devers, P. K. Economic associations among causes of species endangerment in the United States. *Bioscience* **50**, 593–601 (2000).

4. Peterson, M. N., Chen, X. D. & Liu, J. G. Household location choices: Implications for biodiversity conservation. *Conservation Biology* **22**, 912–921 (2008), doi:10.1111/j.1523-1739.2008.00929.x.

5. Peterson, M. N., Chen, X. D. & Liu, J. G. Household location choices.

6. Hardin, G. *Living Within Limits: Ecology, Economics, and Population Taboos.* Oxford University Press, 1993.

7. Peterson, M. N., Peterson, M. J., Peterson, T. R. & Liu, J. G. A household perspective for biodiversity conservation. *Journal of Wildlife Management* **71**, 1243–1248 (2007), doi:10.2193/2006-207.

8. Louv, R. *Last Child in the Woods: Saving Our Children from Nature Deficit Disorder*. Algonquin Books, 2005.

9. Mengel, J. Making Haiti 100% renewable: U.S. company joins with Clinton Global Initiative for ambitious project. *Green Chip Stocks*, October 11 (2010), www.greenchipstocks.com/articles/making-haiti-100-renewable/1127/.

10. Harvey, D. *MegaCities, Lecture 4*. Twynstra Gudde Management Consultants, 2000.

11. Hirschhorn, J. S. Environment, quality of life, and urban growth in the new economy. *Environmental Quality Management* **10**, 1–8 (2001).

12 Rawls, J. *Political Liberalism*. Columbia University Press, 1993.

13. Fishman, R. The fifth migration. *Journal of the American Planning Association* **71**, 357–367 (2005).

14. Malizia, E. Comment on "Planning leadership in a new era." *Journal of the American Planning Association* **72**, 407–409 (2006). Myers, D. & Ryu, S. H. Aging baby boomers and the generational housing bubble: Foresight and mitigation of an epic transition. *Journal of the American Planning Association* **74**, 17–33 (2008).

15. Fishman, R. The fifth migration. Nelson, A. C. Leadership in a new era. *Journal of the American Planning Association* **72**, 393–407 (2006). Myers, D. & Ryu, S. H. Aging baby boomers.